PRAISE FOR NEIL deGRASSE TYSON

"[Tyson] tackles a great range of subjects . . . with great humor, humility, and—most important—humanity."

—*Entertainment Weekly*

"[A] looming figure. . . . [A]n astronomer to the bone."

—Carl Zimmer, *Playboy*

"It's one thing to be a lauded astrophysicist. It's another to possess a gift for comedic timing. You don't normally get both, but that's Neil."

—Jon Stewart, *The Daily Show*

"Tyson is a rock star whose passion for the laws of nature is matched by his engaging explanations of topics ranging from dark matter to the absurdity of zombies."

—*Parade*

"[Tyson] is bursting with ideas."

—Lisa de Moraes, *Washington Post*

"Neil deGrasse Tyson might just be the best spokesperson for science alive."

—Matt Blum, *Wired*

"It's more imperative than ever that we find writers who can explain not only what we're discovering, but how we're discovering it. Neil deGrasse Tyson is one of those writers."

—Anthony Doerr, *Boston Sunday Globe*

"Tyson is a confidently smooth popularizer of science."

—*People*

"The heir-apparent to Carl Sagan's rare combination of wisdom and communicative powers."

—Seth MacFarlane, creator of *Family Guy*

T0036877

ORIGINS

ALSO BY NEIL deGRASSE TYSON

Starry Messenger: Cosmic Perspectives on Civilization

Letters from an Astrophysicist

Astrophysics for People in a Hurry

Astrophysics for Young People in a Hurry

Accessory to War: The Unspoken Alliance between Astrophysics and the Military (with Avis Lang)

Cosmic Queries: StarTalk's Guide to Who We Are, How We Got Here, and Where We're Going

Space Chronicles: Facing the Ultimate Frontier

The Pluto Files: The Rise and Fall of America's Favorite Planet

Death by Black Hole: And Other Cosmic Quandaries

ALSO BY DONALD GOLDSMITH

The End of Astronauts: Why Robots Are the Future of Exploration (with Martin Rees)

Exoplanets: Hidden Worlds and the Quest for Extraterrestrial Life

The Astronomers

The Runaway Universe: The Race to Find the
Future of the Cosmos

Nemesis: The Death-Star and Other Theories of
Mass Extinction

Einstein's Greatest Blunder? The Cosmological Constant
and Other Fudge Factors in the Physics of the Universe

E = Einstein: His Life, His Thought, and His Influence on
Our Culture (with Marcia Bartusiak)

ORIGINS

FOURTEEN BILLION YEARS
OF COSMIC EVOLUTION

REVISED AND UPDATED

NEIL deGRASSE TYSON AND DONALD GOLDSMITH

W. W. NORTON & COMPANY
Independent Publishers Since 1923

For information about permission to reproduce selections from this book, write to Permissions, W. W. Norton & Company, Inc.
500 Fifth Avenue, New York, NY 10110

For information about special discounts for bulk purchases, please contact W. W. Norton Special Sales at specialsales@wwnorton.com or 800-233-4830

Manufacturing by Lakeside Book Company
Book design by Chris Welch
Production manager: Erin Reilly

ISBN 978-0-393-86688-9 (pbk)

W. W. Norton & Company, Inc., 500 Fifth Avenue, New York, N.Y. 10110
www.wwnorton.com

W. W. Norton & Company Ltd., 15 Carlisle Street, London W1D 3BS

1 2 3 4 5 6 7 8 9 0

To all those who look up,

And to all those who do not yet know

why they should

CONTENTS

PART IV: THE ORIGIN OF LIFE

ORIGINS

INTRODUCTION

A MEDITATION ON THE ORIGINS OF SCIENCE AND THE SCIENCE OF ORIGINS

A new synthesis of scientific knowledge has emerged and continues to flourish. In recent years, the answers to questions about our cosmic origins have not come solely from the domain of astrophysics. Working under the umbrella of emergent fields with names such as astrochemistry, astrobiology, and astro-particle physics, astrophysicists have recognized that they can benefit greatly from the collaborative infusion of other sciences. To invoke multiple branches of science when answering the question "Where did we come from?" empowers investigators with a previously unimagined breadth and depth of insight into how the universe works.

In this second edition of *Origins: Fourteen Billion Years of Cosmic Evolution*, we introduce the reader to this new synthesis of knowledge, incorporating new discoveries in biology, astronomy, and astrophysics, including startling results such as these:

- Five thousand newly detected "exoplanets" embody an enormously rich variety of surface conditions and orbital characteristics. Some of their environments strongly favor the origin and existence of life, pointing the way to estimating the abundance of life in the cosmos.

- Astrophysicists now deploy an entirely new class of detectors, capable of responding to the gravitational radiation from violent events, billions of light-years from Earth, that Einstein's theories had predicted but had never been directly recorded until 2017. Observations from three worldwide facilities have revealed, among other marvels, the merger of black holes dozens of times more massive than our Sun.

- Once regarded as far too cold or too small to offer the possibility of harboring life, five celestial bodies other than Mars now qualify as well worth investigating. These include Ceres, the largest asteroid; Jupiter's moon Europa and Saturn's moon Enceladus, with extensive oceans beneath a worldwide ice cover; and Saturn's giant moon Titan, where lakes of liquid nitrogen could play the role that water does on Earth.

- A profusion of new ground-based and spaceborne observatories have sharpened our views of the distant universe, not only in visible light but also in radio, infrared, and other domains. Their improved capabilities have brought forth a discrepancy between two key methods for determining how rapidly the universe is expanding. The currently unknown resolution of this "crisis in cosmology" may lead to new understanding of the physics laws that govern the cosmos.

These and other significant discoveries allow us to address not only the origin of the universe but also the origin of the largest structures that matter has formed, the origin of the stars that light the cosmos, the origin of planets that offer the likeliest sites for life, and the origin of life itself on at least one of those planets and potentially elsewhere in the solar system and throughout the universe.

Humans remain fascinated with the topic of origins for reasons both logical and emotional. We can hardly comprehend the essence of anything without knowing where it came from. And of all the stories that we hear, those that recount our own origins engender the deepest resonance within us.

The self-centeredness bred into our bones by our evolution and experience on Earth has led us naturally to focus on local events and phenomena in the retelling of most origin stories. However, every advance in our knowledge of the cosmos has revealed that we live on a cosmic speck of dust, orbiting a representative star in the far suburbs of a common sort of galaxy, among at least a hundred billion galaxies in the universe. The news of our cosmic unimportance triggers powerful defense mechanisms in the human psyche. Many of us unwittingly resemble the man in the cartoon who gazes at the starry heavens and remarks to his companion, "When I look at all those stars, I'm struck by how insignificant they are."

Throughout history, different cultures have produced creation myths that explain our origins as the result of cosmic forces shaping our destiny. These histories have helped us to ward off feelings of insignificance. Although origin stories typically begin with the big picture, they get down to Earth with impressive speed, zipping past the creation of the universe, of all its contents, and of life on Earth, to arrive at long explanations of complex details of human history and its social conflicts, as if we somehow formed the center of creation.

Almost all the disparate answers to the quest for origins

accept as their underlying premise that the cosmos behaves in accordance with general rules that reveal themselves, at least in principle, to our careful examination of the world around us. Ancient Greek philosophers raised this premise to exalted heights, insisting that we humans possess the power to perceive how nature operates, as well as the underlying reality beneath what we observe: the fundamental truths that govern all else. Quite understandably, they insisted that uncovering those truths would be difficult. Twenty-four centuries ago, in his most famous reflection on our ignorance, the philosopher Plato compared those who strive for knowledge to prisoners chained in a cave, unable to see objects behind them, who must therefore attempt to deduce from the shadows of these objects an accurate description of reality.

With this simile, Plato not only summarized humanity's attempts to understand the cosmos but also emphasized that we have a natural tendency to believe that mysterious, dimly sensed entities govern the universe, privy to knowledge that we can, at best, glimpse only in part. From Plato to Buddha, from Moses to Muhammad, from a hypothesized cosmic creator to modern films about "the matrix," humans in every culture have concluded that higher powers rule the cosmos, gifted with an understanding of the gulf between reality and superficial appearance.

Half a millennium ago, a new attitude toward understanding nature slowly took hold. This approach, which we now call science, arose from the confluence of new technologies and the discoveries that they fostered. The spread of printed books across Europe, together with simultaneous improvements in travel by road and water, allowed individuals to communicate more quickly and effectively, so that they could learn what others had to say and could respond far more rapidly than in the past. During the sixteenth and seventeenth centuries, this hastened back-and-forth disputation and led to a new way of acquiring knowledge, based on the principle that the most effective means of understanding the cosmos relies on careful observations, coupled with attempts to specify broad and basic principles that explain a set of these observations.

One more concept gave birth to science. Science depends on organized skepticism, that is, on continual, methodical doubting. Few of us doubt our own conclusions, so science encourages its skeptical approach by approving those who doubt someone else's. We may rightly call this approach unnatural, not so much because it calls for mistrusting someone else's thoughts, but because science rewards those who can demonstrate that another scientist's conclusions are just plain wrong. To other scientists, the scientist who corrects a colleague's error, or cites good reasons for seriously doubting their conclusions, performs a noble deed, like a Zen master who boxes the ears of a novice straying from the meditative path, although scientists correct one another more as equals than as master and student. By rewarding a scientist who spots another's errors—a task that human nature makes much easier than discerning one's own mistakes—scientists as a group have created an inborn system of self-correction. Scientists have collectively created our most efficient and effective tool for analyzing nature, because they seek to disprove other scientists' theories even as they support their earnest attempts to advance human knowledge. Science therefore forms a collective pursuit without being a mutual admiration society.

Like all attempts at human progress, the scientific approach works better in theory than in practice. Not all scientists doubt one another as effectively as they should. The need to impress

scientists who occupy powerful positions, and who are sometimes swayed by factors that lie beyond their conscious knowledge, can interfere with science's self-correcting ability. In the
long run, however, errors cannot endure, because other scientists will discover them and promote their own careers by trum-
peting the news. Those conclusions that do survive the attacks
of other scientists eventually achieve the status of scientific
"laws," accepted as valid descriptions of reality, even though
scientists understand that each of these laws may someday find
itself to be only part of, and subject to, a larger, deeper truth.

But scientists hardly spend all their time attempting to
prove one another mistaken. Most scientific endeavors proceed by testing imperfectly established hypotheses against
slightly improved observational results. Every once in a while,
however, a significantly novel take on an important theory
emerges, or (more often in an age of technological advances)
an entirely new range of observations opens the way to a
fresh set of hypotheses. The greatest moments in scientific
history have arisen, and will always arise, when an original
explanation, perhaps coupled with new observational results,
produces a seismic shift in our conclusions about the work-
ings of nature. Scientific progress depends on individuals in
both camps: those who assemble better data and extrapolate
carefully from them, and those who risk much—and have
much to gain if successful—by challenging widely accepted
conclusions.

Science's skeptical core makes it a poor competitor for human
hearts and minds, which recoil from ongoing controversies and
much prefer the security of seemingly eternal truths. If the scientific approach were just one more pathway to interpretation
of the cosmos, it would never have amounted to much; but its

big-time success rests on the fact that it works. If you board an aircraft built according to science—with principles that have survived numerous attempts to prove them wrong—you have a far better chance of reaching your destination than you do in an aircraft constructed by the rules of Vedic astrology.

Throughout relatively recent history, people confronted with the success of science in explaining natural phenomena have reacted in one of four ways. First, a small minority have embraced the scientific method as our best hope for understanding nature, and seek no additional ways to comprehend the universe. Second, a much larger number ignore science, judging it uninteresting, opaque, or opposed to the human spirit. (Those who watch television greedily without ever pausing to wonder how the pictures and sound reach them remind us that the words "magic" and "machine" share deep etymological roots.) Another minority, conscious of the assault that science seems to make upon their cherished beliefs, seek actively to disprove scientific results that annoy or enrage them. They do so, however, quite outside the skeptical framework of science, as you can easily establish by asking one of them, "What evidence would convince you that you are wrong?" These anti-scientists still feel the distress that John Donne described in his poem "An Anatomy of the World: The First Anniversary," written in 1611 as the first fruits of modern science appeared:

And new philosophy calls all in doubt,
The element of fire is quite put out,
The Sun is lost, and th'earth, and no man's wit
Can well direct him where to look for it.
And freely men confess that this world's spent,

When in the planets and the firmament
They seek so many new; they see that this [world]
Is crumbled out again to his atomies.
'Tis all in pieces, all coherence gone . . .

Fourth, another large section of the public accept the scientific approach to nature while maintaining a belief in entities existing beyond our complete understanding that rule the cosmos. Baruch Spinoza, the philosopher who created the strongest bridge between the natural and the supernatural, rejected any distinction between nature and God, insisting instead that the cosmos is simultaneously nature and God. Adherents of more conventional religions, which typically insist on this distinction, often reconcile the two by mentally separating the domains in which the natural and the supernatural operate.

No matter what camp you are in, no one doubts that these are auspicious times for learning what's new in the cosmos. Let us, then, proceed with our adventurous quest for cosmic origins, acting much like detectives who deduce the facts of the crime from the evidence left behind. We invite you to join us in search of cosmic clues—and the means of interpreting them—so that together we may uncover the story of how part of the universe turned into ourselves.

PART I
THE ORIGIN OF
THE UNIVERSE

1

IN THE BEGINNING

I n the beginning, there was physics. "Physics" describes how
matter, energy, space, and time behave and interact with
one another. The interplay of these four characters in our
cosmic drama underlies all biological and chemical phenomena. Hence everything fundamental and familiar to us Earth-
lings begins with, and rests upon, the laws of physics. When
we apply these laws to astronomical settings, we deal with
physics writ large, which we call astrophysics.

In almost any area of scientific inquiry, but especially in
physics, the frontiers of discovery live at the extremes of our
ability to measure events and situations. In an extreme of matter, such as the neighborhood of a black hole, gravity strongly
warps the surrounding space-time continuum. At an extreme
of energy, thermonuclear fusion sustains itself within the 15-
million-degree cores of stars. And at every physical extreme
we find the outrageously hot and dense conditions that pre-
vailed during the first few moments of the universe. To understand what happens in each of these scenarios requires laws of
physics discovered after 1900, during what physicists now call

the modern era, to distinguish it from the classical era, which includes all previous physics.

One major feature of classical physics is that events and laws and predictions actually make sense when you stop and think about them. They were all discovered and tested in ordinary laboratories in ordinary buildings. The laws of gravity and motion, of electricity and magnetism, and of the nature and behavior of heat energy are still taught in high school physics classes. These revelations about the natural world fueled the Industrial Revolution, itself transforming culture and society in ways unimagined by generations that came before, and remain central to what happens, and why, in the world of everyday experience.

By contrast, nothing makes sense in modern physics, because everything happens in regimes that lie far beyond those to which our human senses respond. This is a good thing. We may happily report that our daily lives remain wholly devoid of extreme physics. On a normal morning, you get out of bed, wander around the house, eat something, then dash out the front door. At day's end your loved ones fully expect you to look no different than you did when you left, and to return home in one piece. But imagine yourself arriving at the office, walking into an overheated conference room for an important 10 a.m. meeting, and suddenly losing all your electrons—or worse yet, having every atom of your body fly apart. That would be bad. Suppose instead that you're sitting in your office trying to get some work done by the light of your desk lamp, when somebody flicks on a thousand watts of overhead lights, causing your body to bounce randomly from wall to wall until you're jack-in-the-boxed out the window. Or what if you go to a sumo-wrestling match after work, only to see the two nearly spherical gentlemen collide, disappear, and then spontaneously become two beams of light that leave the room in opposite directions? Or suppose that on your way home, you take a road less traveled, and a darkened building sucks you in feet first, stretching your body head to toe while squeezing you shoulder to shoulder as you get extruded through a hole, never to be seen or heard from again. If those scenes played themselves out in our daily lives, we would find modern physics far less bizarre; our knowledge of the foundations of relativity and quantum mechanics would flow naturally from our life experiences; and our loved ones would probably never let us go to work. But back in the early minutes of the universe that kind of stuff happened all the time. To envision it, and to understand it, we have no choice but to revise our notions of common sense and to create an altered intuition about how matter behaves, and how physical laws describe its behavior, at extremes of temperature, density, and pressure.

We must enter the world of $E = mc^2$.

Albert Einstein first published a version of this famous equation in 1905, the year in which his seminal research paper entitled "Zur Elektrodynamik bewegter Körper" appeared in *Annalen der Physik*, the preeminent German journal of physics. The paper's title in English reads "On the Electrodynamics of Moving Bodies," but the work is far better known as Einstein's special theory of relativity, which introduced concepts that forever changed our notions of space and time. Just 26 years old, working as a patent examiner in Bern, Switzerland, Einstein offered further details, including his best-known equation, in another, remarkably short (two-and-a-half-page) paper published later the same year in the same journal: "Ist die Trägheit eines Körpers von seinem Energieinhalt abhängig?," or "Does the Inertia of a Body Depend on Its Energy Content?" To save you the effort of locating the original article, of designing an experiment, and of thus testing Einstein's theory, the answer to the paper's title is yes. As Einstein wrote, using L where today we employ E, and V for the speed of light where today we employ c:

> If a body gives off the energy L in the form of radiation, its mass diminishes by L/V^2. . . . The mass of a body is a measure of its energy-content; if the energy changes by L, the mass changes in the same sense by $L/9 \times 10^{20}$, the energy being measured in ergs, and the mass in grams.

Considering, as scientists do, how to verify the truth of his statement, he then suggested,

> It is not impossible that with bodies whose energy-content is variable to a high degree (e.g. with radium salts) the theory may be successfully put to the test.[*]

There you have it: the algebraic recipe for all occasions when you want to convert matter into energy, or energy into matter. $E = mc^2$—energy equals mass times the square of the speed of light—gives us a supremely powerful computational tool that extends our capacity to know and understand the universe from as it is now, all the way back to infinitesimal fractions of a second after the birth of the cosmos. With this equation, you

---

[*] Albert Einstein, *The Principle of Relativity*, trans. W. Perrett and G. B. Jeffrey (London: Methuen and Company, 1923), 69–71.

can tell how much radiant energy a star can produce, or how much you could gain by converting the coins in your pocket into useful forms of energy.

The most familiar form of energy—shining all around us, though often unrecognized and unnamed in our mind's eye—is the photon, a massless, irreducible particle of visible light or of any other form of electromagnetic radiation. We all live within a continuous bath of photons: from the Sun, the Moon, and the stars; from your stove, your chandelier, and your night-light; from hundreds of radio and television stations; and from countless cellphone and radar transmissions. Why, then, don't we actually see the daily transmuting of energy into matter, or of matter into energy? The energy of common photons sits far below the mass of the least massive subatomic particles, when converted into energy by $E = mc^2$. Because these photons wield too little energy to become anything else, they lead sim-ple, relatively uneventful lives. They also remind us that Ein-stein's formula references the energy locked within particles with mass, and not the energy of motion, or "kinetic energy," the only form of energy that photons carry.

Do you long for some action with $E = mc^2$? Start hanging around gamma-ray photons that have some real energy—at least 200,000 times more than visible photons. You'll quickly get sick and die of the cancer induced by these high-energy particles; but before that happens, you'll see pairs of electrons, one made of matter, the other of antimatter (just one of many dynamic particle-antiparticle duos in the universe), pop into existence where photons once roamed. As you watch, you'll also see matter-antimatter pairs of electrons collide, annihi-lating each other and creating gamma-ray photons once again. Increase the photons' energy by another factor of 2,000, and

you now have gamma rays with enough energy to turn suscep-
tible people into the Hulk. Pairs of these photons wield enough
energy, fully described by the power of $E = mc^2$, to create particles such as neutrons, protons, and their antimatter partners, each with nearly 2,000 times the mass of an electron.
High-energy photons don't hang out just anywhere, but they do
exist in many a cosmic crucible. For gamma rays, almost any
environment hotter than a few billion degrees will do just fine.

The cosmological significance of particles and energy packets that transform themselves into one another is staggering.
Currently, the temperature of our expanding universe, found
by measuring the bath of microwave photons that pervades
all of space, is a mere 2.73 on the Kelvin scale. (In the Kelvin
system, whose intervals are denoted by K and equal those of
the Centigrade scale, all temperatures are positive: at absolute
zero, 0 in the Kelvin system, particles have the least possible
energy; room temperature is about 295 K; and water boils at
373 K.) Like the photons of visible light, microwave photons
are too cool to have any realistic ambitions of turning themselves into particles via $E = mc^2$. In other words, no known
particle has a mass so low that it can be made from the mea-
ger energy of a microwave photon. The same holds true for the
photons that form radio waves, infrared, and visible light, as
well as ultraviolet and X-rays. More simply expressed, all particle transmutations require gamma rays. Yesterday, however,
the universe was a little bit smaller and a little bit hotter than
today. The day before, it was smaller and hotter still. Roll the
clocks backward some more—say, 13.8 billion years—and you
land squarely in the post–big bang primordial soup, a time
when the temperature of the cosmos was high enough to be
astrophysically interesting, as gamma rays filled the universe.

To understand the behavior of space, time, matter, and energy from the big bang to present day is one of the greatest triumphs of human thought. If you seek a complete explanation for the events of the earliest moments, when the universe was smaller and hotter than ever thereafter, you must find a way to enable the four known forces of nature—gravity, electromagnetism, and the strong and the weak nuclear forces—to talk to one another, to unify and become a single meta-force. You must also find a way to reconcile two currently incompatible branches of physics: quantum mechanics (the science of the small) and general relativity (the science of the large).

———

Spurred by the successful marriage of quantum mechanics and electromagnetism during the mid-twentieth century, just as electricity and magnetism had been triumphantly integrated the century before, physicists moved swiftly to blend quantum mechanics and general relativity into a single and coherent theory: quantum gravity. Although so far they have all failed, we already know where the high hurdles lie: during the "Planck era." That's the cosmic phase up to 10^{-43} second (one ten-million-trillion-trillion-trillionth of a second) after the beginning. Because information can never travel more rapidly than the speed of light, 3×10^8 meters per second, a hypothetical observer situated anywhere in the universe during the Planck era could see no farther than 3×10^{-35} meter (30 trillion-trillion-trillionths of a meter). The German physicist Max Planck, after whom these unimaginably small times and distances are named, introduced the idea of "quantized energy"—the concept that energy appears only in discrete packets—in 1900, and generally receives credit as the father of quantum mechanics.

Not to worry, though, so far as daily life goes. The clash between quantum mechanics and gravity poses no practical problem for the contemporary universe. Astrophysicists apply the tenets and tools of general relativity and quantum mechanics to extremely different classes of problems. But in the beginning, during the Planck era, the large was small, so there must have been a kind of shotgun wedding between the two. Alas, the vows exchanged during that ceremony continue to elude us, so no (known) laws of physics describe with any confidence how the universe behaved during the brief honeymoon, before the expanding universe forced the very large and very small to part ways.

At the end of the Planck era, gravity wriggled itself loose from the other, still-unified forces of nature, achieving an independent identity nicely described by our current theories. As the universe aged past 10^{-35} second, it continued to expand and to cool, and what remained of the once-unified forces divided into the electroweak force and the strong nuclear force. Later still, the electroweak force split into the electromagnetic and the weak nuclear forces, laying bare four distinct and familiar forces—with the weak force controlling radioactive decay, the strong force binding together the particles in each atomic nucleus, the electromagnetic force holding atoms together in molecules, and gravity binding matter in bulk. By the time the universe aged a trillionth of a second, its transmogrified forces, along with other critical transfigurations, had already imbued the cosmos with its fundamental properties, each worthy of its own book.

While time dragged on for the universe's first trillionth of a second, the interplay of matter and energy continued incessantly. Shortly before, during, and after the strong and electroweak forces had split, the universe contained a seething ocean of quarks, leptons, and their antimatter siblings, along with bosons, the particles that enable particles to interact with one another. None of these particle families, so far as we now know, can be divided into anything smaller or more basic. Fundamental though they are, each family of particles comes in several species. Photons, including those that form visible light, belong to the boson family. The leptons most familiar to the nonphysicist are electrons and perhaps neutrinos; and the most familiar quarks are . . . well, there are no familiar quarks, because in ordinary life we always find quarks bound together into larger particles such as protons and neutrons. Each species of quark has been assigned an abstract name that serves no real philological, philosophical, or pedagogical purpose except to distinguish it from the others: "up" and "down," "strange" and "charmed," and "top" and "bottom." Bosons, by the way, derive their name from the Indian physicist Satyendra Nath Bose. The word "lepton" comes from the Greek *leptos*, meaning "light" or "small." "Quark," however, has a literary and far more imaginative origin. The American physicist Murray Gell-Mann, who in 1964 proposed the existence of quarks, and who then thought that the quark family had only three members, drew the name from a characteristically elusive line in James Joyce's *Finnegans Wake*: "Three quarks for Muster Mark!" One advantage quarks can claim: all their names are simple—something that chemists, biologists, and geologists seem unable to achieve in naming their own stuff.

Quarks are quirky. Unlike protons, which each have an electric charge of +1, and electrons, each with a charge of –1, quarks have fractional charges that come in units of ⅓. And except under the most extreme conditions, you'll never catch a quark all by itself; it will always be clutching on to one or two other quarks. In fact, the force that keeps two (or more) of them together actually grows *stronger* as you separate them—as if some sort of subnuclear rubber band held them together. Separate the quarks sufficiently far, and the rubber band snaps. The energy stored in the stretched band now summons $E = mc^2$ to create a new quark at each end, leaving you back where you started.

During the quark-lepton era in the cosmos's first trillionth of a second, the universe had a density sufficient for the average separation between unattached quarks to rival the separation between attached quarks. Under those conditions, allegiances between adjacent quarks could not be established unambiguously, so they moved freely among themselves. The experimental detection of this state of matter as a mixture of quarks alone, understandably named "quark soup," was reported for the first time in 2002 by a team of physicists working at the Brookhaven National Laboratory on Long Island.

The combination of observation and theory suggests that an episode in the very early universe, perhaps during one of the splits between different types of force, endowed the cosmos with a remarkable asymmetry, in which particles of matter outnumbered their antimatter counterparts by only about one part in a billion—a difference that allows us to exist today. That tiny discrepancy in population could hardly have been noticed amid the continuous creation, annihilation, and recreation of quarks and antiquarks, electrons and anti-electrons (better known as positrons), and neutrinos and antineutrinos. During that era, the slight preponderance of matter over antimatter had plenty of opportunities to find other particles with which to annihilate, and so did all the other particles.

But not for much longer. As the universe continued to expand and cool, its temperature fell rapidly below 1 trillion kelvins. A millionth of a second had now passed since the beginning, but this tepid universe no longer had a temperature or density sufficient to cook quarks. All the quarks quickly grabbed dance partners, creating a permanent new family of heavy particles called hadrons (from the Greek *hadros*, meaning "thick"). That quark-to-hadron transition quickly produced protons and neutrons as well as other, less familiar types of heavy particles, all composed of various combinations of quarks. The slight matter-antimatter asymmetry in the quark-lepton soup now passed to the hadrons, with extraordinary consequences.

As the universe cooled, the amount of energy available for the spontaneous creation of particles declined continuously. During the hadron era, photons could no longer invoke $E = mc^2$ to manufacture quark-antiquark pairs: their E could not cover the pairs' mc^2. In addition, the photons that emerged from all the remaining annihilations continued to lose energy to the ever-expanding universe, so their energies eventually fell below the threshold required to create hadron-antihadron pairs. Every billion annihilations left a billion photons in their wake—and only a single hadron survived, mute testimony to the tiny excess of matter over antimatter in the early universe. Those lone hadrons would ultimately get to have all the fun that matter can enjoy: they would provide the source of galaxies, stars, planets, and people.

Without the imbalance of a billion and one to a mere billion between matter and antimatter particles, all the mass in the universe (except for the dark matter, whose form remains unknown) would have annihilated before the universe's first second had passed, leaving a cosmos in which we could see (if we had existed) photons *and nothing else*—the ultimate Let-there-be-light scenario.

By now, one second of time has passed.

At 1 billion degrees, the universe remains piping hot—still able to cook electrons, which, along with their positron (antimatter) counterparts, continue to pop in and out of existence. But within the ever-expanding, ever-cooling universe, their days (seconds, really) are numbered. What was formerly true for hadrons now comes true for electrons and positrons: they annihilate each other, and only one electron in a billion emerges, the lone survivor of the matter-antimatter suicide pact. The other electrons and positrons died to flood the universe with a greater sea of photons.

With the era of electron-positron annihilation over, the cosmos has "frozen" into existence one electron for every proton. As the cosmos continues to cool, with its temperature falling below 100 million degrees, its protons fuse with other protons and with neutrons, forming atomic nuclei and hatching a universe in which 90 percent of these nuclei are hydrogen and 10 percent are helium, along with relatively tiny numbers of deuterium, tritium, and lithium nuclei.

Two minutes have now passed since the beginning.

Not for another 380,000 years does much happen to our particle soup of hydrogen nuclei, helium nuclei, electrons, and photons. Throughout these hundreds of millennia, the cosmic temperature remains sufficiently hot for the electrons to roam free among the photons, batting them to and fro.

As we will shortly detail in Chapter 3, this freedom comes to an abrupt end when the temperature of the universe falls below 3,000 K (about half the temperature of the Sun's surface). Right about then, all the electrons acquire orbits around the nuclei, and thereby form atoms. The marriage of electrons with nuclei leaves the newly formed atoms within a ubiquitous bath of visible light photons, completing the story of how particles and atoms formed in the primordial universe.

As the universe continues to expand, its photons continue to lose energy. Today, in every direction astrophysicists look, they find a cosmic fingerprint of microwave photons at a temperature of 2.73 K, which represents a thousandfold decline in the photons' energies since the time atoms first formed. The photons' patterns on the sky—the exact amount of energy that arrives from different directions—retain a memory of the cosmic distribution of matter just before atoms formed. From these patterns, astrophysicists can obtain remarkable knowledge, including the age and shape of the universe. Even though atoms now form part of daily life in the universe, Einstein's equation still has plenty of work to do—in particle accelerators, where matter-antimatter particle pairs are created routinely from energy fields; in the core of the Sun, where 4.4 million tons of matter are converted into energy every second; and in the cores of all other stars.

$E = mc^2$ also manages to apply itself near black holes, just outside their event horizons, where particle-antiparticle pairs can pop into existence at the expense of the black hole's formidable gravitational energy. The British cosmologist Stephen Hawking first described the hijinks in 1975, showing that the entire mass of a black hole can slowly evaporate by this mechanism. In other words, black holes are not entirely black. The phenomenon is known as Hawking radiation, and serves as a reminder of the continued fertility of Einstein's most famous equation.

But what happened *before* all this cosmic fury? What happened before the beginning?

Astrophysicists have no idea. Rather, our most creative ideas have little or no grounding in experimental science. Yet the religious faithful tend to assert, often with a tinge of smugness, that something must have started it all: a force greater than all others, a source from which everything issues. A prime mover. In the mind of such a person that something is, of course, God, whose nature varies from believer to believer but who always bears the responsibility for starting the ball rolling.

But what if the universe was always there, in a state or condition that we have yet to identify—a multiverse, for example, in which everything we call the universe amounts to only a tiny bubble in an ocean of suds? Or what if the universe, like its particles, just popped into existence from nothing we could see?

These rejoinders typically satisfy no one. Nonetheless, they remind us that informed ignorance provides the natural state of mind for research scientists at the ever-shifting frontiers of knowledge. People who believe themselves ignorant of nothing have neither looked for, nor stumbled upon, the boundary between what is known and unknown in the cosmos. And therein lies a fascinating dichotomy. "The universe always was" gets no respect as a legitimate answer to "What was around before the beginning?" But for many religious people, the answer "God always was" is the obvious and pleasing answer to "What was around before God?"

No matter who you may be, engaging yourself in the quest to discover where and how everything began can induce emotional fervor—as if knowing our beginnings would bestow upon you some form of fellowship with, or perhaps governance over, all that comes later. So what is true for life itself is true for the universe: knowing where you came from is just as important as knowing where you are going.

2

ANTIMATTER MATTERS

Particle physicists have won the contest for the most peculiar, yet playful, jargon of all the physical sciences. In the preceding chapter, we met not only protons, neutrons, and electrons but also photons, hadrons, bosons, and quarks. But this hardly plumbs the depths of particle-physics names. Where else in science could you find a neutral vector boson exchanged between a negative muon and a muon neutrino? Or a gluon exchange binding together a strange quark and a charmed quark? And where else can you meet squarks, photinos, and gravitinos?

Alongside these seemingly countless particles with peculiar names, particle physicists must contend with the parallel existence of *anti*particles, collectively known as antimatter. Don't be fooled by its persistent appearance in science-fiction stories: antimatter is real. And as you might have heard, an antimatter particle does annihilate upon contact with its ordinary-matter counterpart.

The universe reveals a peculiar romance between antiparticles and particles. They can be born together out of pure energy, and they can annihilate as they reconvert their combined mass back into pure energy. Antimatter can pop into existence out of thin air—or rather, thin space. Gamma-ray photons with sufficiently high energy can transform themselves into electron-positron pairs, thus converting all of their seriously large energy into a small amount of matter, in a process whose energy budget fulfills $E = mc^2$. The energy embodied in the mass of the electron-positron pair comes from the energy of motion embodied in the gamma-ray photon.

In 1932, the American physicist Carl David Anderson discovered the anti-electron, the positively charged, antimatter counterpart to the negatively charged electron. Since then, particle physicists have routinely made antiparticle varieties in the world's particle accelerators, and have even assembled antiparticles into whole atoms. Since 1996, an international group led by Walter Oelert of the Institute for Nuclear Physics Research in Jülich, Germany, has created atoms of antihydrogen, in which an anti-electron calmly orbits an antiproton. To make these first anti-atoms, the physicists used the giant particle accelerator operated by the European Organization for Nuclear Research (better known by its French acronym CERN) in Geneva, Switzerland, where so many important contributions to particle physics have occurred.

The physicists use a simple creation method: make a bunch of anti-electrons and a bunch of antiprotons, bring them together at a suitable temperature and density, and wait for them to combine to form atoms. During their first round of experiments, Oelert's team produced nine atoms of antihydrogen. But in a world dominated by ordinary matter, life as an antimatter atom can be precarious. The antihydrogen atoms survived for less than 40 nanoseconds (40 billionths of a second) before annihilating with ordinary atoms.

The discovery of the anti-electron was one of the great triumphs of theoretical physics, for its existence had been predicted, just a few years earlier, in an equation created by the British-born physicist Paul A. M. Dirac to describe elementary particles. To describe matter on the smallest size scales—those of atomic and subatomic particles—physicists developed a new branch of physics during the 1920s to explain the results of their experiments with these particles. Using these newly established rules, now known as quantum theory or quantum mechanics, Dirac postulated from a second solution to his equation that a phantom electron from the "other side" might occasionally pop into the world as an ordinary electron, leaving behind a gap or hole in the sea of negative energies. Although Dirac hoped to explain protons in this way, other physicists suggested that this hole would reveal itself experimentally as a positively charged electron—an "anti-electron"—which had come to be known as a positron for its positive electric charge. The detection of actual positrons confirmed Dirac's basic insight and established antimatter as worthy of as much respect as matter itself.

Equations with double solutions are not unusual. One of the simplest examples answers the question, What number times itself equals nine? Is it 3 or −3? Of course, the answer is both, because $3 \times 3 = 9$ and $−3 \times −3 = 9$. Physicists cannot guarantee that all the solutions of an equation correspond to events in the real world, but if a mathematical model of a physical phenomenon is correct, manipulating its equations can be as useful as (and somewhat easier than) manipulating the entire universe. As with Dirac and antimatter, such steps often lead to verifiable predictions. If the predictions prove incorrect, then the theory is discarded. But regardless of the physical outcome, a mathematical model ensures that the conclusions you may draw from it are both logical and internally consistent.

Subatomic particles have many measurable features, of which mass and electric charge rank among the most important. Except for the particle's mass, which is always the same for a particle and its antiparticle, the specific properties of each type of antiparticle will always be precisely opposite to those of the "non-anti-" version of the particle. For example, the positron has the same mass as the electron, but the positron has one unit of positive charge while the electron has one unit of negative charge. Similarly, the antiproton provides the oppositely charged antiparticle of the proton.

Believe it or not, the chargeless neutron also has an antiparticle. It's called—you guessed it—the antineutron. An antineutron has an opposite zero charge with respect to the ordinary neutron. This arithmetical magic derives from the particular triplet of fractionally charged particles (the quarks) that form neutrons. The three quarks that compose a neutron have charges of $-\frac{1}{3}$, $-\frac{1}{3}$, and $+\frac{2}{3}$, while those in the antineutron have charges of $+\frac{1}{3}$, $+\frac{1}{3}$, and $-\frac{2}{3}$. Each set of three quark charges adds up to a net charge of zero, though the corresponding components do have opposite charges.

Antimatter can pop into existence out of thin air. If gamma-ray photons have sufficiently high energy, they can transform themselves into electron-positron pairs, thus converting all of their seriously large energy into a small amount of matter, in a process whose energy side fulfills Einstein's famous equation $E = mc^2$. In the language of Dirac's original interpretation, the gamma-ray photon kicked an electron out of the domain of negative energies, creating an ordinary electron and an electron hole. The reverse process can also occur. If a particle and an antiparticle collide, they will annihilate by refilling the hole and turning into gamma rays. As we have stressed, gamma rays are the sort of high-energy radiation you should avoid.

If you could somehow manage to manufacture a blob of anti-particles at home, you would have a wolf by the ears. Storage would immediately become a challenge, because your anti-particles would annihilate with the material of any conventional sack or grocery bag (either paper or plastic) in which you chose to confine or carry them. A cleverer storage mechanism involves trapping the charged antiparticles within the confines of a strong magnetic field, where they are repelled by invisible but highly effective magnetic "walls." If you embed the magnetic field in a vacuum, you can render the antiparticles safe from annihilation with ordinary matter. This magnetic equivalent of a bottle will also be the bag of choice whenever you must handle other container-hostile materials, such as the 100-million-degree glowing gases involved in (controlled) nuclear fusion experiments. The greatest storage problem arises after you have created entire anti-atoms, because anti-atoms, like atoms, do not normally rebound from a magnetic wall. You would be wise to keep your positrons and antiprotons in separate magnetic bottles until you must bring them together.

To generate antimatter—for example, from high-energy gamma-ray photons—requires at least as much energy as you can recover when the antimatter annihilates with matter to become pure energy again. Unless you had a full tank of antimatter fuel before launch, a self-generating antimatter engine would slowly suck energy from your starship. Perhaps the original *Star Trek* television and film series embodied this fact, but if memory serves, Captain Kirk continually asked for "more power" from the matter-antimatter drives, and Scotty warned in his Scottish accent that "the engines cannot take it."

Although physicists expect hydrogen and antihydrogen atoms to behave identically, they have not yet verified this prediction experimentally, mainly because of the difficulty they face in keeping antihydrogen atoms in existence, rather than having them annihilate almost immediately from contact with protons and electrons. Physicists would like to verify that the detailed behavior of a positron bound to an antiproton in an antihydrogen atom obeys all the laws of quantum theory, and that an anti-atom's gravity behaves precisely as we expect of ordinary atoms. Could an anti-atom produce antigravity (repulsive) instead of ordinary gravity (attractive)? All theory points toward the latter, but the former, if it should prove correct, would offer amazing new insights into nature. On atomic-sized scales, the force of gravity between any two particles is immeasurably small. Instead of gravity, electromagnetic and nuclear forces dominate the behavior of these tiny particles, because both forces are much, much stronger than gravity. To test for antigravity, you would need enough anti-atoms to make ordinary-sized objects, so that you can measure their bulk properties and compare them to ordinary matter. If a set of billiard balls (and, of course, the billiard table and the cue sticks) were made of antimatter, would a game of anti-pool be indistinguishable from a game of pool? Would an anti–eight ball fall into the corner pocket in exactly the same way as an ordinary eight ball? Would anti-planets orbit an anti-star in the same way that ordinary planets orbit ordinary stars?

It's philosophically sensible, and in line with all the predictions of modern physics, to presume that the bulk properties of antimatter will prove to be identical to those of ordinary matter—normal gravity, normal collisions, normal light, and so forth. Unfortunately, this means that if an anti-galaxy were headed our way, on a collision course with our Milky Way, it would remain indistinguishable from an ordinary galaxy until it was too late to do anything about it. But this fearsome fate cannot be common in the universe today because if, for example, a single anti-star annihilated with a single ordinary star, the conversion of their matter and antimatter into gamma-ray energy would be swift, violent, and total. If two such stars with masses similar to the Sun's (each containing 10^{57} particles) were to collide in our galaxy, their melding would produce an object so luminous that it would temporarily outproduce all the energy of all the stars of 100 million galaxies and fry us to an untimely end. We have no compelling evidence that such an event has ever occurred anywhere in the universe. So, as best we can judge, the universe is dominated by ordinary matter, and has been since the first few minutes after the big bang. Thus total annihilation through matter-antimatter collisions need not rank among your chief safety concerns on your next intergalactic voyage.

Still, the universe now seems disturbingly imbalanced: theories of the early moments of the universe suggest that particles and antiparticles should have been created in equal numbers, yet we find a cosmos dominated by ordinary particles, which seem to be perfectly happy without their antiparticles. Do hidden pockets of antimatter in the universe account for the imbalance? Was a law of physics violated (or was an unknown law of physics at work?) during the early universe, forever tipping the balance in favor of matter over antimatter? Recent, tantalizing results from CERN have hinted that antimatter, when left alone long enough, may spontaneously convert to matter—violating all known laws of particle physics. We may never fully know the answers to these questions. But for now, if an alien hovers over your front lawn and extends an appendage as a gesture of greeting, toss it your eight ball before you get too friendly. If the appendage and the ball explode, the alien probably consists of antimatter. (How the alien's followers will react to this result, and what the explosion will do to you, need not detain us here.) If nothing untoward occurs, you may safely take your new friend to your leader.

3

LET THE LIGHT SHINE

A glance at a clear dark sky confirms that the universe sparkles with light. At night we can see the light from some of the Sun's closest neighbors, and can marvel in the knowledge that beyond the stars visible to unaided eyes lie hundreds of billions of other stars in our Milky Way, plus trillions and trillions more in faraway galaxies. The modern development of our cosmological knowledge, however, rests not on visible light observations but on the ever-enlarging use of light's cousins, which, while remaining invisible to humans, reach us laden with information.

The light that we see occupies a small, central portion of the entire spectrum of electromagnetic radiation, which stretches from the shortest-wavelength gamma rays at one end to the longest-wavelength radio waves at the other. Each type of electromagnetic radiation consists of photons, massless particles that all move through space at the same speed, the "speed of light," covering about 186,000 miles, or 300,000 kilometers, every second. Photons differ in their wavelengths, in their frequencies of vibration, and in the energy that each photon carries. Einstein's famous formula describes the energy con-

tained in a particle's mass, of which photons possess not a
whit. They do, however, carry energy of motion. This allows
them to affect matter: for example, photons of visible light can
induce chemical changes in the retinas of your eyes. Gamma
rays, with the largest amount of energy per photon, represent
a danger to human tissue; radio waves, with the smallest, can
pass through walls (and us) without significant interaction.

If we like, we can refer to all photons as "light," always bearing in mind that the full spectrum of "light" includes many
varieties. This terminology offers an excellent reminder of the
fundamental similarity among all photon types, and underlies
a poetic description of the cosmos: the universe was born in a
blaze of light that filled all of space, continues to do so, and
always will do so. Since then, the ongoing expansion of the
universe has steadily stretched the photons' wavelengths, and
diluted their energy. Fourteen billion years later, the blaze of
light has become a glow so modest that it passed by, utterly
undetected, until 1964.

During the earliest moments after the big bang, the universe
contained a frothing mass of enormously energetic particles,
whose violent collisions produced and destroyed every type of
particle and antiparticle almost instantaneously. But as the
universe expanded, the creation of new space reduced the particles' energies. Half an hour after the big bang, the universal
epoch of creation and destruction came to an end. By this time,
the cosmos had established its basic mix of "ordinary" matter—familiar types of matter, to be contrasted with the mysterious "dark matter" that we discuss in Chapter 4. Ordinary
matter appeared in only a few dominant varieties: protons,

electrons, helium nuclei (each made of two protons and two neutrons), a flood of photons, and an equally impressive flood of ghostly particles called neutrinos.

For the next 3,800 centuries, while the universe's expansion continued to reduce the energy of the frothing particles, not much else changed. The universe remained opaque to photons, which could travel only tiny distances before meeting a freely moving electron and bouncing off in a new direction. If your mission had been to see across the universe, you couldn't have done so. Any photons heading toward your eye would, just nano-seconds or picoseconds earlier, have bounced off electrons right in front of your face to produce a glowing fog in all directions. Electrons remained unattached because impacts from the sea of photons promptly interfered with their natural tendency to create atoms by entering orbits around protons or helium nuclei. All newly created atoms met immediate destruction when energetic photons struck them, knocking their electrons loose again. These constant interactions between photons and matter smoothed the universe, so that every cubic centimeter had almost identical densities of matter, the same numbers of photons, and the same temperature—everywhere.

Cosmologists characterize a sea of photons such as those that fill the universe by a descriptive temperature. Any object not at absolute zero (and, let's face it, no objects have that temperature) will radiate photons of various energies, but will produce the most photons at an energy that depends on its temperature. Your body, for example, with a temperature of about 310 degrees above absolute zero, radiates primarily infrared photons. Scientists specify this temperature as 310 K, where K denotes the Kelvin absolute temperature scale, which begins with 0 at absolute zero, and uses the same temperature intervals as the Celsius scale. Much hotter objects—the hotter stars, for instance—radiate most of their photons in the energies that characterize visible light. The temperature in a group of particles with mass, measured in kelvins, varies in direct proportion to the average kinetic energy of the particle, and vice versa. The peak in energy of the radiation that the particles produce can be specified by the particles' temperature. When astrophysicists say, for example, that radiation from the Sun has a characteristic temperature of 6,000 K, they imply that the gas releasing this radiation has that temperature.

The creation of new space—an occurrence far easier to write down than to conceive, but one that nonetheless took place throughout past time and continues to occur now—dilutes all particles' energies. Eventually, *none* of the photons had energies sufficient to free electrons from their atomic prisons upon impact. The photons, energy-deprived as they were, found themselves free to roam the cosmos at the speed of light. The "time of decoupling," 380,000 years after the big bang, stands out as one of the key markers in cosmic history: the transition from an opaque to a transparent universe. Ever since then, the photons that filled (and still fill) the cosmos have traveled without hindrance through space, changed only by the steady reduction in their energy produced by the universe's expansion. Having begun as gamma-ray photons, the photons morphed into ultraviolet, visible light, and infrared photons. As their wavelengths grew larger, they became steadily less energetic, but they never stopped being photons.

Astrophysicists coined the phrase "cosmic background radiation" to describe the universal sea of photons released to roam free after atoms could form and persist throughout the universe. Abbreviated as the "CBR," today, 13.8 billion years after the beginning, the photons have shifted across the spectrum to become microwaves. For this reason, astrophysicists sometimes call the CBR the "cosmic microwave background." One hundred billion years from now, when the universe will have expanded and cooled considerably more, the photons in the radiation will have still lower energies, so astrophysicists of the far future may describe the CBR as the "cosmic radio-wave background."

When we observe the CBR, we study photons that have traveled for nearly 14 billion years, and we know that any photons created here before the time of decoupling have likewise traveled far from us at the speed of light for the same amount of time. The fact that the CBR arrives in nearly the same quantity from all directions implies that the universe was homogeneous, nearly the same everywhere.

Why should we care about this radiation? What can this cosmic sea of photons do for us? The answer, which embodies enormous informational though little practical importance, rests on the fact that these photons carry the fingerprint of a long-vanished past, the furthest back in time that we can hope to observe (barring advances that may await us in the twenty-second century), and reveal critical facts about the young universe, when it was one forty-thousandth of its present age.

Those critical facts—the details that bring devilish satisfaction to astrophysicists—reside in the tiny *differences* in the numbers and energies of the photons in the cosmic background radiation that arrive from different directions. These variations arise from deviations from perfect smoothness in the distribution of matter at the time of decoupling. Some regions had a density of matter slightly greater than average, some a bit less. All the structure in the universe today arose from these density differences, recorded for all time in the sea of photons. Regions with higher densities had a better chance to form enormous clusters of galaxies; those with lower densities were left bereft of the chance to concentrate matter, and became the voids between galaxy clusters.

The CBR provides an excellent example of how the competition among rival theories will eventually yield a winner when sufficiently accurate observations become available. Its discovery belongs to a notable class: those that scientists predicted before they observed them—and in this case, the prediction was made two decades before the technology that could prove it came into existence. In 1927, the Belgian Catholic priest Georges Lemaître, who was also a cosmologist (this makes a certain sense, of course), drew on Albert Einstein's theory of general relativity to create the concept of a "primeval atom"— basically a precursor of the big bang model of the universe. Twenty years later, following Lemaître's line of reasoning, the Ukrainian-born physicist George Gamow (by then an American citizen), in collaboration with Ralph Alpher and Robert Herman, concluded that the early universe must have been immensely hot, and then cooled steadily as time went on. Alpher and Herman proceeded to use the laws of physics to describe the expansion of the universe after the time of decoupling, when atoms first formed and photons could roam freely, and concluded that the CBR should now have a temperature close to 5 K.

Yes, they got the number wrong—the CBR actually has a temperature of 2.73 K. But these three physicists nevertheless performed a successful extrapolation back into the depths of long-vanished cosmic epochs—as great a feat as any other in the history of science. To take some basic atomic physics from a slab in the lab, and to deduce from it the largest-scale phenomenon ever measured—the temperature history of our universe—ranks as nothing short of mind-blowing. Assessing this accomplishment, J. Richard Gott III, an astrophysicist at Princeton University, wrote in *Time Travel in Einstein's Universe*: "Predicting that the radiation existed and then getting its temperature correct to within a factor of 2 was a remarkable accomplishment—rather like predicting that a flying saucer 50 feet in width would land on the White House lawn and then watching one 27 feet in width actually show up."

When Gamow, Alpher, and Herman made their predictions, physicists were still undecided about the story of how the universe began. In 1948, the same year that Alpher and Herman's paper appeared, a rival "steady state" theory of the universe appeared in two papers published in England, one coauthored by the mathematician Hermann Bondi and the astrophysicist Thomas Gold, the other by the cosmologist Fred Hoyle. The steady state theory requires that the universe, though always expanding, has always looked the same—a hypothesis with a deeply attractive simplicity. Because the universe is expanding, and because a steady state universe would not have been any hotter or denser yesterday than today, the Bondi-Gold-Hoyle scenario maintained that matter continuously pops into our universe at just the right rate to maintain a constant average density in the expanding cosmos. In contrast, the big bang theory (given its name in scorn by Fred Hoyle) requires that all matter come into existence at one instant, which some find more emotionally satisfying. The steady state theory takes the issue of the origin of the universe and throws it backward an infinite distance in time—highly convenient for those who would rather not examine this thorny problem.

The prediction of the cosmic background radiation amounted to a shot across the bow of the steady state theorists. The CBR's existence as the product of the earliest epochs of the cosmos would clearly demonstrate that the universe was once far different—much denser and hotter—from the way we find it today. The first direct observations of the CBR therefore put the first nails in the coffin of the steady state theory (though Fred Hoyle never fully accepted the CBR as disproving his elegant theory, going to his grave attempting to explain the radiation as arising from other causes). The cosmic refutation of the steady state theory appeared in 1964, when the CBR was inadvertently and serendipitously discovered by Arno Penzias and Robert Wilson at the Bell Telephone Laboratories (Bell Labs, for short) headquartered in Murray Hill, New Jersey. Their good luck and hard work brought Penzias and Wilson the Nobel Prize just over a decade later.

What led Penzias and Wilson to their Nobel Prize–winning discovery? During the early 1960s, physicists all knew about microwaves, but almost no one had created the capability of detecting weak signals in the microwave portion of the spectrum. Back then, most wireless communication (e.g., receivers, detectors, and transmitters) rode on radio waves, which have longer wavelengths than microwaves. For microwaves, scientists needed a shorter-wavelength detector and a sensitive antenna to capture them. Bell Labs had one, a king-size, horn-shaped antenna on Crawford Hill, the highest point (380 feet!) on New Jersey's coastal plain, that could focus and detect microwaves as well as any apparatus on Earth.

If you're going to send or receive a signal of any kind, you don't want other signals to contaminate it. Penzias and Wilson were trying to open up a new channel of communication for Bell Labs, so they wanted to pin down the amount of contaminating "background" interference these signals would experience—from the Sun, from the center of our galaxy, from terrestrial sources, from whatever. They embarked on a standard, important, and entirely innocent set of measurements, aimed at establishing how easily they could detect microwave signals. Though Penzias and Wilson had some astronomy background, they were not cosmologists but technophysicists studying microwaves, unaware of the predictions made by Gamow, Alpher, and Herman. What they were decidedly *not* looking for was the cosmic microwave background.

So they ran their experiment, and corrected their data for all known sources of interference. When they found background noise in the signal that didn't go away, they couldn't figure out how to get rid of it. The noise seemed to come from every direction above the horizon, and it didn't change with time. Finally they looked inside their giant horn. Pigeons were nesting there, leaving a white dielectric substance (pigeon poop) everywhere nearby. Things must have been getting desperate for Penzias and Wilson: Could the droppings, they wondered, be responsible for the background noise? They cleaned it up, and sure enough, the noise dropped a bit. But it still wouldn't go away. The paper they published in 1965 in the *Astrophysical Journal* refers to the persistent puzzle of an inexplicable "excess antenna temperature," rather than the astronomical discovery of the century.

While Penzias and Wilson were scrubbing bird droppings from their antenna, a team of physicists led by Robert H. Dicke at Princeton University was building a detector specifically designed to find the CBR that Gamow, Alpher, and Herman had predicted. Lacking the resources of Bell Labs, the professors' work proceeded more slowly. The moment that Dicke and his colleagues heard about Penzias and Wilson's results, they knew that they'd been scooped. The Princeton team understood exactly what the "excess antenna temperature" was. Everything fit the theory: the temperature, the fact that the signal came from all directions in equal amounts, and that it wasn't linked in time with Earth's rotation or Earth's position in orbit around the Sun.

But why should this interpretation gain general acceptance? For good reason. Photons take time to reach us from distant parts of the cosmos, so we inevitably look back in time whenever we look outward into space. This means that if the intelligent inhabitants of a galaxy far, far away measured the temperature of the cosmic background radiation for themselves, long before we managed to do so, they should have found its temperature to be greater than 2.73 K, because they would have inhabited the universe when it was younger, smaller, and hotter than it is today.

Can such an audacious assertion be tested? Yup. Turns out that the compound of carbon and nitrogen called cyanogen—dangerous on Earth though prevalent in the universe—will become excited by exposure to microwaves. If the microwaves are warmer than the ones in our CBR, they will excite the molecule a little more effectively than our microwaves do. The cyanogen compounds thus act as a cosmic thermometer. When we observe them in distant galaxies, which we see when they were significantly younger than the universe today, these mol-

ecules should have found themselves bathed in a warmer cosmic background than the cyanogen in our Milky Way galaxy.
In other words, those galaxies ought to live more excited lives
than we do. And they do. The spectrum of cyanogen in distant
galaxies shows that the microwaves had just the temperature
that we expect at these earlier cosmic times.

You would need a seriously impressive imagination to make
this stuff up.

The CBR does far more for astrophysicists than provide direct
evidence for a hot early universe and thus for the big bang
model. It turns out that the photons that compose the CBR
reach us bursting with information about the cosmos both
before and after the universe became transparent. We have
noted that until that time, about 380,000 years after the big
bang, the universe was opaque, so you couldn't have witnessed
matter making shapes even if you'd been sitting front-row center. You couldn't have seen where galaxy clusters were starting
to form. Before anybody, anywhere, could see anything worth
seeing, photons had to acquire the ability to travel, unimpeded,
across the universe. When the time was right, each photon
began its cross-cosmos journey at the point where it smacked
into the last electron that would ever stand in its way. As more
and more photons escaped without being deflected by electrons
(thanks to electrons joining nuclei to form atoms), they created an expanding shell of photons that astrophysicists call
"the surface of last scatter." That shell, which formed during
a period of about a hundred thousand years, marks the epoch
when almost all the atoms in the cosmos were born.

By then, matter in large regions of the universe had already begun to coalesce. Where matter accumulates, gravity grows stronger, enabling more and more matter to gather. Those matter-rich regions seeded the formation of galaxy superclusters, while other regions remained relatively empty. The photons that last scattered off electrons within the coalescing regions developed a different, slightly cooler spectrum as they climbed out of the strengthening gravity field, which robbed them of a bit of energy.

The CBR indeed shows spots that are slightly hotter or slightly cooler than average, typically by about one hundred-thousandth of a degree. These hot and cool spots mark the earliest structures in the cosmos, the first clumping together of matter. We know what matter looks like in the universe today because we see galaxies, galaxy clusters, and galaxy superclusters. To figure out how those systems arose, we probe the cosmic background radiation, a remarkable relic from the remote past, still filling the entire universe. Studying the patterns in the CBR amounts to a kind of cosmic phrenology: we can read the bumps on the "skull" of the youthful universe and from them deduce behavior not only for its infancy but also for its grown-up state.

By adding other observations of the local and the distant universe, astrophysicists can determine all kinds of fundamental cosmic properties from the CBR. Compare the distribution of sizes and temperatures of the slightly warmer and cooler areas, for instance, and we can infer the strength of gravity in the early universe, and thus how quickly matter accumulated. From that we can then deduce how much ordinary matter, dark matter, and dark energy the universe comprises (the percentages are 5, 27, and 68, respectively). From there, it's easy to tell whether or not the universe will expand forever, and whether or not the expansion will slow down or speed up as time passes.

Ordinary matter is what everyone is made of. It exerts gravity and can absorb, emit, and otherwise interact with light. Dark matter, as we'll see in Chapter 4, is a substance of unknown nature that produces gravity but does not interact with light in any known way. And dark energy, as we'll see in Chapter 5, induces an acceleration of the cosmic expansion, forcing the universe to expand more rapidly than it otherwise would. The net effect of our cosmic phrenology exam now implies that cosmologists understand how the early universe behaved, but that most of the universe, then and now, consists of stuff they're clueless about. Profound areas of ignorance notwithstanding, today, as never before, cosmology has an anchor. The CBR carries the imprint of a portal through which we all once passed along with everything else in the universe.

The discovery of the cosmic microwave background added new precision to cosmology by verifying the conclusion, originally derived from observations of distant galaxies, that the universe has been expanding for billions of years. It was the accurate and detailed map of the CBR—a map first made for small patches of the sky using balloon-borne instruments and a telescope at the South Pole, and then over the entire sky by a satellite called the Wilkinson Microwave Anisotropy Probe (WMAP)—that secured cosmology's place at the table of experimental science. WMAP's successor, the European Space Agency's Planck satellite, made even higher-resolution observations of the CBR, adding more detail that would prove useful in deducing the parameters that describe the early universe.

Cosmologists have plenty of ego: How else could they have the audacity to infer what brought the universe into being?

But the new era of observational cosmology may call for a more modest, less freewheeling stance among its practitioners. Each new observation, each morsel of data, can be good or bad for your theories. On the one hand, the observations provide a basic foundation for cosmology, a foundation that so many other sciences can take for granted because they achieve rich streams of laboratory observations. On the other hand, new data will almost certainly dispatch some of the tall tales that theorists dreamt up when they lacked the observations that would give them thumbs up or down.

But no science achieves maturity without precision data. Six decades after Penzias and Wilson's discovery, new generations of spacecraft have transformed cosmology into precision science.

4

LET THERE BE DARK

Gravity, the most familiar of nature's forces, is simultaneously the best- and the least-understood phenomenon in nature. It required the mind of Isaac Newton, the millennium's most brilliant and influential, to realize that gravity's mysterious "action at a distance" arises from the natural effects of every bit of matter, and that the attractive forces between any two objects can be described by a simple algebraic equation. It took the mind of Albert Einstein, the twentieth century's most brilliant and influential, to show that we can more accurately describe gravity's action-at-a-distance as a warp in the fabric of space-time, produced by any combination of matter and energy. Einstein demonstrated that Newton's theory requires some modification to describe gravity accurately—in predicting, for example, the amount by which light rays will bend when they pass by a massive object. Although Einstein's equations are fancier than Newton's, they nicely accommodate the matter that we have come to know and love. Matter that we can see, touch, feel, and occasionally taste.

We've now been waiting for nearly a century for another

scientist with a brilliant mind to explain an unexpected and amazing result: the bulk of all the gravitational forces that we've measured in the universe arise from substances that we have neither seen, nor touched, nor felt, nor tasted. Maybe the excess gravity doesn't come from matter at all, but emanates from some other conceptual thing. In any case, we are without a clue. We find ourselves no closer to an answer today than we were when this "missing mass" problem was first identified in 1933 by astrophysicists who measured the velocities of galaxies affected by the gravity of their close neighbors, and was then more fully analyzed in 1937 by the colorful Bulgarian-Swiss-American astrophysicist Fritz Zwicky, who taught at the California Institute of Technology for more than 40 years, presenting his far-ranging insights into the cosmos with an animated means of expression and an impressive ability to antagonize his colleagues.

Zwicky studied the movement of galaxies within an enormous cluster of galaxies, located far beyond the local stars of the Milky Way, that delineate the constellation Coma Berenices (the "hair of Berenice," an Egyptian queen in antiquity). The Coma cluster, as it is called within the profession, is an isolated and richly populated ensemble of galaxies about 325 million light-years from Earth. Its many thousands of galaxies orbit the cluster's center, moving in all directions like bees circling their hive. Studying the motions of a few dozen galaxies to trace the gravity field that binds the entire cluster, Zwicky discovered their average velocity to be shockingly high. Since larger gravitational forces induce higher velocities in the objects that they attract, Zwicky deduced an enormous mass for the Coma cluster. When we sum up all of its galaxies' estimated masses, Coma ranks among the largest and most massive galaxy clusters in the universe. Even so, the cluster does not contain enough visible matter to account for the observed speeds of its member galaxies. Matter seems to be missing.

If you apply Newton's law of gravity, and assume that the cluster does not exist in an odd state of expansion or collapse, you can calculate what the characteristic average speed of its galaxies ought to be. All you need is the size of the cluster and an estimate of its total mass: the mass, acting over distances characterized by the cluster's size, determines how rapidly the galaxies must move to avoid falling into the cluster's center or escaping from the cluster entirely.

In a similar calculation, as Newton showed, you can derive the speed at which each of the planets must move in its orbit at its particular distance from the Sun. Far from being magic, these speeds satisfy the gravitational circumstance in which each planet finds itself. If the Sun suddenly acquired more mass, Earth and everything else in the solar system would require higher velocities to stay in their current orbits instead of falling inward toward the Sun. With too much speed, however, the Sun's gravity would be insufficient to maintain everybody's orbit. If Earth's orbital speed were more than the square root of 2 times its current speed, our planet would achieve "escape velocity" and, you guessed it, escape the solar system. We can apply the same reasoning to much larger objects, such as our own Milky Way galaxy, in which stars move in orbits that respond to the gravity from all the other stars, or in clusters of galaxies, where each of the galaxies likewise responds to the gravity from all the other galaxies. As Einstein once

wrote (more ringingly in German than in this English transla-
tion) to honor Isaac Newton:

Look unto the stars to teach us
How the master's thoughts can reach us
Each one follows Newton's math
Silently along its path.

When we examine the Coma cluster, as Zwicky did during
the 1930s, we find that its member galaxies all move more rapidly than the escape velocity for the cluster, once we establish
that velocity by summing all the galaxy masses taken one by
one, which we estimate from the galaxies' brightnesses. The
cluster should therefore swiftly fly apart, leaving barely a trace
of its beehive existence after just a few hundred million years,
perhaps a billion, had passed. But the cluster is more than 10
billion years old, nearly as old as the universe itself. And so was
born what remains the longest-standing mystery in astronomy.
     Through the decades that followed Zwicky's work, other gal-
axy clusters revealed the same problem. Coma could not be
blamed as simply an oddity. Then whom should we blame?
Newton? No, his theories had been examined for 250 years and
passed all tests. Einstein? No. The formidable gravity of galaxy clusters does not rise high enough to require the full ham-
mer of Einstein's general theory of relativity, just two decades
old when Zwicky published his analysis. Perhaps the "missing
mass" needed to bind the Coma cluster's galaxies does exist,
but in some unknown, invisible form. For a time, astrophysicists called this problem the "missing-light problem," since
they had accurately inferred the cluster's mass from the excess
of gravity and concluded that some process had inhibited the

cluster from producing the corresponding amount of light. Today, with better determinations of the properties of galaxies and galaxy clusters, astrophysicists use the moniker "dark matter," although "dark gravity" would be still more precise.

The dark matter problem reared its invisible head a generation later. In 1976, Vera Rubin, an astrophysicist at the Carnegie Institution of Washington, discovered a similar "missing-mass" anomaly within spiral galaxies themselves. Studying the speeds at which stars orbit their galaxy centers, Rubin first found what she expected: within the visible disk of each galaxy, the stars farther from the center move at greater speeds than stars close in. The farther stars have more matter (stars and gas) between themselves and the galaxy center, requiring higher speeds to sustain their orbits. Beyond the galaxy's luminous disk, however, we can still find some isolated gas clouds and a few bright stars. Using these objects as tracers of the gravity field "outside" the galaxy, where visible matter no longer adds to the total, Rubin discovered that their orbital speeds, which should have fallen with increasing distance out there in Nowheresville, in fact remained high.

These largely empty volumes of space—the rural regions of each galaxy—contain too little visible matter to explain the orbital speeds of the tracers. Rubin correctly reasoned that some form of dark matter must lie in these far-out regions, well beyond the visible edge of each spiral galaxy. Indeed, the dark matter forms a kind of halo around the entire galaxy.

This halo problem exists under our noses, right in our own Milky Way galaxy. From galaxy to galaxy and from cluster to cluster, the discrepancy between the mass in visible objects and the total mass of systems ranges from a factor of just two or three up to factors of many hundreds. Across the universe, the factor averages to about six. That is, cosmic dark matter enjoys about six times the mass of all the visible matter.

Almost five decades later, further research has revealed that most of the dark matter cannot consist of nonluminous ordinary matter. This conclusion rests on two lines of reasoning. First, we can eliminate with near certainty all plausible familiar candidates, like the suspects in a police lineup: Could the dark matter reside in black holes? No, we think that we would have detected this many black holes from their gravitational effects on nearby stars. Could it be dark clouds? No, they would absorb or otherwise interact with light from stars behind them, which real dark matter doesn't do. Could it be interstellar (or intergalactic) planets, asteroids, and comets, all of which produce no light of their own? It's hard to believe that the universe would manufacture six times as much mass in planets as in stars. That would mean six thousand Jupiters for every star in the galaxy, or worse yet, two million Earths. In our own solar system, for example, everything that is not the Sun sums to a paltry 0.2 percent of the Sun's mass.

Thus, as best we can figure, the dark matter doesn't simply consist of matter that happens to be dark. Instead, it's something else altogether. Dark matter exerts gravity according to the same rules that ordinary matter follows, but it does little else that might allow us to detect it. Of course, we are hamstrung in this analysis by not knowing what the dark matter is. The difficulties of detecting dark matter, intimately connected with our difficulties in perceiving what it might be, raise the question: If all matter has mass, and all mass has gravity, does all gravity have matter? We don't know. The name "dark matter" presupposes the existence of a kind of matter that has gravity and that we don't yet understand. But maybe it's the gravity that we don't understand. To study dark matter beyond deducing its existence, astrophysicists now seek to learn where the stuff collects in space. If dark matter existed only at the outer edges of galaxy clusters, for example, then the galaxies' velocities would show no evidence of a dark matter problem, because only sources of gravity *interior* to the galaxies' orbits will affect their speeds. Considered all together, the gravitational forces from sources beyond the galaxies' orbits cancel one another out because they act in opposing directions. If the dark matter occupied only the clusters' centers, then the run of galaxy speeds as measured from the center of the cluster out to its edge would respond to ordinary matter alone. But the speeds of galaxies in clusters reveal that the dark matter permeates the entire volume occupied by the orbiting galaxies. In fact, the locations of ordinary matter and dark matter essentially coincide, and the vast spaces between galaxies have comparatively little dark matter. Several decades ago, a team led by the American astrophysicist J. Anthony Tyson, then at Bell Labs and now at UC Davis (he's called "Cousin Tony" by one of us, though we have no family relationship) produced the first detailed map of the distribution of dark matter's gravity in and around a huge cluster of galaxies. Wherever we see big galaxies, we also find a higher concentration of dark matter within the cluster. The converse is also true: regions with no visible galaxies have a dearth of dark matter.

The discrepancy between dark and ordinary matter varies significantly from one astrophysical environment to another, but it becomes most pronounced for large entities such as galaxies and galaxy clusters. For the smallest objects, such as moons and planets, no discrepancy exists. Earth's surface gravity, for example, can be explained entirely by what's under our feet. So if you are overweight on Earth, don't blame dark matter. Dark matter also has no bearing on the Moon's orbit around Earth, nor on the movements of the planets around the Sun. But we do need it to explain the motions of stars around the center of the galaxy.

Does a different kind of gravitational physics operate on the galactic scale? Probably not, though the possibility cannot be ruled out. More likely, dark matter consists of matter whose nature we have yet to divine, and which clusters more diffusely than ordinary matter does. Otherwise, we would find that one in every six pieces of dark matter has a chunk of ordinary matter clinging to it. So far as we can tell, that's not the way things are.

At the risk of inducing depression, or at least cosmic humility, astrophysicists sometimes argue that all the matter that we have come to know and love in the universe—the stuff of stars, planets, and life—are mere markers adrift in a vast cosmic ocean of something that looks like nothing. But what if this conclusion were entirely wrong? When nothing else seems to work, some scientists will understandably, and quite rightly, question the fundamental laws of physics that underlie the assumptions made by others who seek to understand the universe.

During the early 1980s, the Israeli physicist Mordehai Milgrom of the Weizmann Institute of Science in Rehovot, Israel, suggested a change in Newton's laws of gravity, a theory now known as MOND (MOdified Newtonian Dynamics). Accepting

the fact that standard Newtonian dynamics operates successfully on size scales smaller than galaxies, Milgrom suggested
that Newton needed some help in describing gravity's effects
at distances the sizes of galaxies and galaxy clusters, within
which individual stars and star clusters are so far apart that
they exert relatively little gravitational force on each other.
Milgrom added an extra term to Newton's equation, specifically tailored to come to life at astronomically large distances.
Although he invented MOND as a computational tool, Milgrom didn't rule out the possibility that his theory could refer
to a new phenomenon of nature.

MOND has had only limited success. The theory can account
for the movement of isolated objects in the outer reaches of
many spiral galaxies, but it raises more questions than it
answers. MOND fails to predict reliably the dynamics of more
complex configurations, such as the movement of galaxies in
binary and multiple systems. Furthermore, the detailed map
of the cosmic background radiation produced by the WMAP
satellite in 2003 allowed cosmologists to isolate and measure
the influence of dark matter in the early universe. Because
these results appeared to correspond to a consistent model of
the cosmos based on conventional theories of gravity, MOND
lost many adherents. This problem with MOND grew still
larger when the more precise measurements of the cosmic
background radiation made by the Planck spacecraft rolled in
over the course of the following decade.

During the first half-million years after the big bang, a mere
moment in the 14-billion-year sweep of cosmic history, matter
in the universe had already begun to coalesce into the blobs
that would become clusters and superclusters of galaxies. But
the cosmos was expanding all along, and would double in size

during its next half-million years. So the universe responds to two competing effects: gravity wants to make stuff coagulate, but the expansion wants to dilute it. If you do the math, you rapidly deduce that the gravity from ordinary matter could not win this battle by itself. It needed the help of dark matter, without which we would be living—actually, not living—in a universe with no structure: no clusters, no galaxies, no stars, no planets, no people. How much gravity from dark matter did it need? Six times as much as that provided by ordinary matter itself. This analysis leaves no room for MOND's little corrective terms in Newton's laws. The analysis doesn't tell us what dark matter is, only that dark matter's effects are real. Try as you may, you cannot credit ordinary matter for any of it.

Dark matter plays another crucial role in the universe. To appreciate all that the dark matter has done for us, go back in time to a couple of minutes after the big bang, when the universe was still so immensely hot and dense that hydrogen nuclei (protons) could fuse together. This crucible of the early cosmos forged hydrogen into helium, along with trace amounts of lithium, plus an even smaller amount of deuterium, which is a heavier version of the hydrogen nucleus, with a neutron added to the proton. This mixture of nuclei provides another cosmic fingerprint of the big bang, a relic that allows us to reconstruct what happened when the cosmos was a few minutes old. In creating this fingerprint, the prime mover was the strong nuclear force—the force that binds protons and neutrons within the nucleus—and not gravity, a force so weak that it gains significance only as particles assemble themselves by the trillions.

By the time the temperature dropped below a threshold value, nuclear fusion throughout the universe had created one helium nucleus for every ten hydrogen nuclei. The universe had also turned about one part in a thousand of its ordinary matter into lithium nuclei, and two parts in a hundred thousand into deuterium. If dark matter consisted not of some noninteracting substance but instead of dark ordinary matter—matter with normal fusion privileges—then, because the dark matter packed six times as many particles into the tiny volumes of the early universe as ordinary matter did, its presence would have dramatically increased the fusion rate of hydrogen. The result would have been a noticeable overproduction of helium, in comparison with the observed amount, and the birth of a universe notably different from the one that we inhabit.

Helium is one tough nucleus, relatively easy to make but extremely difficult to fuse into other nuclei. Because stars have continued to make helium from hydrogen in their cores, while destroying relatively little helium through more advanced nuclear fusion, we may expect that the places where we find the lowest amounts of helium in the universe should have no less helium than what the universe produced during its first few minutes. Sure enough, galaxies whose stars have only minimally processed their ingredients show that 1 in 10 of their atoms consists of helium, just as you would expect from the big bang birthday suit of the cosmos, so long as the dark matter then present did not participate in the nuclear fusion that created nuclei.

———

So, dark matter is our friend. But astrophysicists understandably grow uncomfortable whenever they must base their calculations on concepts they don't understand, even though this wouldn't be the first time they've done so. Astrophysicists measured the energy output of the Sun, for instance, long before anybody knew that thermonuclear fusion was responsible. Back in the nineteenth century, before the introduction of quantum mechanics and the discovery of other deep insights into the behavior of matter on its smallest scales, fusion didn't even exist as a concept.

Unrelenting skeptics might compare the dark matter of today with the hypothetical, now-defunct "ether," proposed centuries ago as the weightless, transparent medium through which light moved. For many years, until a famous 1887 experiment in Cleveland performed by Albert Michelson and Edward Morley, physicists assumed that the ether must exist, even though not a shred of evidence supported this presumption. Known to be a wave, light was believed to require something through which to move, much as sound waves move through air. Light turns out to be quite happy, however, to travel through the vacuum of space, devoid of any rippling medium. Unlike sound waves, light waves propagate themselves.

But dark matter ignorance differs fundamentally from ether ignorance. While the ether amounted to a placeholder for our incomplete understanding, the existence of dark matter derives not from mere presumption but from the observed effects of its gravity on visible matter. We're not inventing dark matter out of thin space; instead, we deduce its existence from observational facts. Dark matter is just as real as the thousand planets in orbit around stars other than the Sun that have been discovered solely by their gravitational influence on their host stars. The worst that can happen is that physicists (or others of deep insight) might discover that the dark matter consists not of matter at all but instead of something else, yet they cannot argue it away. Could dark matter be the manifestation of forces from another dimension? Or of a parallel universe intersecting ours? Even so, none of this would change the successful invocation of dark matter's gravity in the equations that we use to understand the formation and evolution of the universe.

Other unrelenting skeptics might declare that "seeing is believing." A seeing-is-believing approach to life works well in many endeavors, including mechanical engineering, fishing, and perhaps dating. It's also good, apparently, for residents of Missouri. But it doesn't make for good science. Science is not just about seeing. Science is about measuring—preferably with something that's *not* your own eyes, which are inextricably conjoined with the baggage of your brain: preconceived ideas, post-conceived notions, imagination unchecked by reference to other data, and bias.

Having resisted attempts to detect it directly on Earth for three quarters of a century, dark matter has become a type of Rorschach test of the investigator. Some particle physicists say the dark matter must consist of some ghostly class of undiscovered particles that interact with matter via gravity, but otherwise interact with matter or light only weakly, or not at all. This sounds off the wall, but the suggestion has precedent. Neutrinos, for instance, are well known to exist, though they interact extremely weakly with ordinary light and matter. Neutrinos from the Sun—two neutrinos are produced for every helium nucleus made in the solar core—travel through the vacuum of space at nearly the speed of light, and pass through Earth as though it does not exist. The tally: night and day, 100 billion neutrinos from the Sun enter, then exit, each square inch of your body every second.

But neutrinos can be stopped. Every rare once in a while they interact with matter via nature's weak nuclear force. And if you can stop a particle, you can detect it. Compare neutrinos' elusive behavior with that of the Invisible Man (in his invisible phase)—as good a candidate for dark matter as anything else. He could walk through walls and doors as though they were not there. Although equipped with these talents, why didn't he just drop through the floor into the basement?

If we can build sufficiently sensitive detectors, the particle physicist's dark matter particles may reveal themselves through familiar interactions. Or they may reveal their presence through forces other than the strong nuclear force, the weak nuclear force, and electromagnetism. These three forces (plus gravity) mediate all interactions between and among all known particles. So the choices are clear. Either dark matter particles must wait for us to discover and to control a new force or class of forces through which the particles interact, or else dark matter particles interact via normal forces, but with staggering weakness.

MOND theorists see no exotic particles in their Rorschach tests. They think gravity, not particles, is what needs fixing. And so they brought forth modified Newtonian dynamics—a bold attempt that seems to have failed but is doubtless the precursor of other efforts to change our view of gravity rather than our census of subatomic particles.

Other physicists pursue what they call TOEs, or "theories of everything." In a spin-off of one version, our own universe indeed lies near a parallel universe, with which we interact only through gravity. You'll never run into any matter from that parallel universe, but you might feel its tug, crossing into the spatial dimensions of our own universe. Imagine a phantom universe right next to ours, revealed to us only through its gravity. Sounds exotic and unbelievable, but probably not any more so than the first suggestions that Earth orbits the Sun, or that our galaxy is not alone in the universe.

So, dark matter's effects are real. We just don't know what the dark matter is. It seems not to interact through the strong force, so it cannot make nuclei. It hasn't been found to interact through the weak nuclear force, something even elusive neutrinos do. It doesn't seem to interact through the electromagnetic force, so it doesn't make molecules, or absorb or emit or reflect or scatter light. It does exert gravity, however, to which ordinary matter responds. That's it. After all these years of investigation, astrophysicists haven't discovered it doing anything else.

Detailed maps of the cosmic background radiation have demonstrated that dark matter must have existed during the first 380,000 years of the universe. We also need dark matter today in our own galaxy and in galaxy clusters to explain the motions of objects they contain. But as far as we know, the march of astrophysics has not yet been derailed or stymied by our ignorance. We simply carry dark matter along as a strange friend, and invoke it where and when the universe requires it of us.

In what we hope is the not-so-distant future, the fun will continue as we learn to exploit dark matter—once we figure out what the stuff is made of. Imagine invisible toys, cars that pass through one another, or superstealth airplanes. The history of obscure and obtuse discoveries in science is rich with examples of clever people who came later and who figured out how to exploit such knowledge for their own gain or for the benefit of life on Earth.

5

LET THERE BE MORE DARK

The cosmos, we now know, has both a light and a dark side. The light side includes all familiar heavenly objects—the stars, which group by the billions into galaxies, as well as the planets and smaller cosmic debris that may not produce visible light but do emit other forms of electromagnetic radiation, such as infrared or radio waves.

We have discovered that the dark side of the universe embraces the puzzling dark matter, detected by its gravitational effects on visible matter, but of completely unknown form and composition. An extremely modest amount of this dark matter may be ordinary matter that remains invisible because it produces no detectable radiation. But, as we explored in the previous chapter, the great bulk of the dark matter must consist of non-ordinary matter that refuses to interact with ordinary matter aside from its gravitational force, so its nature continues to elude us.

Beyond all issues concerning dark matter, the dark side of the universe has another, entirely different aspect. One that resides not in matter of any kind, but in space itself. We owe this concept, along with the amazing results that it implies, to the father of modern cosmology, none other than Albert Einstein himself.

More than a century ago, while the newly perfected machine guns of World War I slaughtered soldiers by the thousands a few hundred miles to the west, Albert Einstein sat in his office in Berlin, pondering the universe. As the war began, Einstein and a colleague had circulated an antiwar petition among their peers, gathering two other signatures in addition to their own. This act set him apart from his fellow scientists, most of whom had signed an appeal in support of Germany's war effort, and ruined his colleague's career. But Einstein's engaging personality and scientific fame allowed him to keep the esteem of his peers. He continued his efforts to find equations that could accurately describe the cosmos.

Before the war ended, Einstein achieved success—arguably his greatest triumph. In November 1915, he produced his general theory of relativity, which describes how space and matter interact: matter tells space how to bend, and space tells matter how to move. To replace Isaac Newton's mysterious "action at a distance," Einstein regarded gravity as a local warp in the fabric of space. The Sun, for example, creates a sort of dimple that bends space most noticeably at distances closest to it. The planets tend to roll into this dimple, but their inertia keeps them from falling all the way in. Instead, they move in orbits around the Sun that keep them at a nearly constant distance from the center of the dimple in space. Within a few weeks after Einstein published his theory, the physicist Karl Schwarzschild, diverting himself from the horrors of life in the German army (which gave him a fatal disease soon afterward), used Einstein's concept to demonstrate that an object with sufficiently strong gravity will create a "singularity" in space. At such a singularity, space bends completely around the object and prevents anything, including light, from leaving its immediate vicinity. We now call these objects black holes.

Einstein's theory of general relativity led him to the key equation he had been seeking, one that links the contents of space to its overall behavior. Studying this equation in the privacy of his office, creating models of the cosmos in his mind and on paper, Einstein almost discovered the expanding universe, a dozen years before observations by Edwin Hubble in California revealed it.

Einstein's basic equation predicts that the space within a universe in which matter has a roughly even distribution cannot be "static." The cosmos cannot just "sit there," as our intuition insists that it should, and as all astronomical observations until that time implied. Instead, the totality of space must always be either expanding or contracting: space must behave something like the surface of an inflating or deflating balloon, but never like the surface of a balloon with constant size.

This worried Einstein. For once, this bold theorist, who mistrusted authority and had never hesitated to oppose conventional physics ideas, felt that he had gone too far. No astronomical observations suggested an expanding or contracting universe, because astrophysicists had only documented the motions of nearby stars and had not yet determined the far-away distances to what we now call galaxies. Rather than announcing to the world that the universe must either be expanding or contracting, Einstein returned to his equation, seeking a change that would immobilize the cosmos.

He soon found one. Einstein's basic equation allowed for a term with a constant but unknown value that represents the amount of energy contained in every cubic centimeter of empty space. Because nothing suggested that this constant term should have one value or another, in his first pass Einstein had set it equal to zero. Now Einstein published a scientific article to demonstrate that if this constant term, which cosmologists later named the "cosmological constant," had a particular value, then space could be static. Then theory would no longer conflict with observations of the universe, and Einstein could regard his equation as valid.

But Einstein's solution encountered grave difficulties. In 1922, a Russian mathematician named Alexander Friedmann proved that Einstein's static universe must be unstable, like a pencil balanced on its point. The slightest ripple or disturbance would cause space either to expand or to contract. Einstein first proclaimed Friedmann mistaken, but then, in a generous act typical of the man, published an article retracting that claim and pronouncing Friedmann correct after all. As the 1920s ended, Einstein was delighted to learn of Hubble's discovery that the universe is expanding. According to George Gamow's recollections, Einstein pronounced the cosmological constant his "greatest blunder." Except for a few cosmologists who continued to invoke a non-zero cosmological constant (with a value different from the one that Einstein had used) to explain certain puzzling observations, most of which later proved to be incorrect, scientists the world over sighed with relief that space had proved to have no need of this constant.

Or so they thought. The great cosmological story at the end of the twentieth century, the surprise that stood the world of cosmology on one ear and sang a different tune into the other, resides in the stunning discovery, first announced in 1998, that the universe does have a non-zero cosmological constant. Empty space does indeed contain energy, now named "dark energy," and possesses highly unusual characteristics that determine the future of the entire universe.

To understand, and possibly even to believe, these dramatic assertions, we must follow the crucial themes in cosmologists' thinking during the seventy years following Hubble's discovery of the expanding universe. Einstein's fundamental equation allows for the possibility that space can have curvature, described mathematically as positive, zero, or negative. Zero curvature describes "flat space," the kind that our minds insist must be the only possibility, which extends to infinity in all directions, like the surface of an infinite chalkboard. In contrast, a positively curved space corresponds in analogy to the surface of a sphere, a two-dimensional space whose curvature we can see by using the third dimension. Notice that the center of the sphere, the point that appears to remain stationary as its two-dimensional surface expands or contracts, resides in this third dimension and appears nowhere on the surface that represents all of space.

Just as all positively curved surfaces include only a finite area, all positively curved spaces contain only a finite volume. A positively curved cosmos has the property that if you journey outward from Earth for a sufficiently long time, you will eventually return to your point of origin, like Magellan circumnavigating our globe. But unlike positively curved spherical surfaces, negatively curved spaces extend to infinity, even though they are not flat. A negatively curved two-dimensional surface resembles the surface of an infinitely large saddle: it curves "upward" in one direction (front to back) and "downward" in another (side to side).

If the cosmological constant equals zero, we can describe the overall properties of the universe with just two numbers. One of these, called the Hubble constant, measures the rate at which the universe is expanding now. The other measures the curvature of space. During the second half of the twentieth century, almost all cosmologists believed that the cosmological constant was zero, and considered measurement of the cosmic expansion rate and the curvature of space as their primary research agenda.

Both of these numbers can be found from accurate measurements of the speeds at which objects located at different distances are receding from us. The overall trend between distance and velocity—the rate at which the recession velocities of galaxies increase with increasing distance—yields the Hubble constant, whereas small deviations from this general trend, which appear only when we observe objects at the greatest distances from us, will reveal the curvature of space. Whenever astrophysicists observe objects many billion light-years from the Milky Way, they look so far back in time that they see the cosmos not as it is now but as it was when significantly less time had elapsed since the big bang. Observations of galaxies located 5 billion or more light-years from the Milky Way allow cosmologists to reconstruct a significant part of the history of the expanding universe. In particular, they can see how the rate of expansion has changed with time—the key to determining the curvature of space. This approach works, at least in principle, because the degree of space's curvature induces subtle differences in the rate at which the universal expansion has changed through the past billions of years.

In practice, astrophysicists remained unable to fulfill this program, because they could not make sufficiently reliable estimates of the distances to galaxy clusters many billion light-years from Earth. They had another arrow in their quiver, however. If they could measure the average density of all the matter in the universe—that is, the average number of grams of material per cubic centimeter of space—they could compare this number with the "critical density," a value predicted by Einstein's equations that describe the expanding universe. The critical density specifies the precise density of matter that exists in a universe with zero curvature of space. If the actual density lies above this value, the universe has positive curvature. In that case, assuming that the cosmological constant equals zero, the cosmos will eventually cease expanding and start contracting. If, however, the actual density exactly equals the critical density, or falls below it, then the universe will expand forever. Exact equality of the actual and critical values of the density occurs in a cosmos with zero curvature, whereas in a negatively curved universe, the actual density is less than the critical density.

By the mid-1990s, cosmologists knew that even after including all the dark matter they had detected from its gravitational influence on visible matter, the total density of matter in the universe came to only about one-quarter of the critical density. This result hardly seems astounding, although it does imply that the cosmos will never cease expanding, and that the space in which all of us live must be negatively curved. It pained theoretically oriented cosmologists, however, because most of them had come to believe that space must have zero curvature.

This belief rested on the "inflationary model" of the universe, named (unsurprisingly) during an era with a steeply rising consumer price index. In 1979, Alan Guth, a physicist working at the Stanford Linear Accelerator Center in California, hypothesized that during its earliest moments, the cosmos expanded at an incredibly rapid rate—so rapidly that different bits of matter accelerated away from one another at speeds far greater than the speed of light. But doesn't Einstein's theory of special relativity make the speed of light a universal speed limit for all motion? Not exactly. Einstein's limit applies only to objects moving *within* space and not to the expansion of space itself. During the "inflationary epoch," which lasted only from about 10^{-37} second to 10^{-34} second after the big bang, the cosmos expanded by a factor of at least 10^{50}.

What produced this enormous cosmic expansion? Guth speculated that all of space must have undergone a "phase transition," a cosmic change analogous to what happens when liquid water quickly freezes into ice. After some crucial tweaking by his colleagues in the Soviet Union, the United Kingdom, and the United States, Guth's idea became so attractive that it has dominated theoretical models of the extremely early universe for the past four decades.

And what makes inflation such an attractive theory? The inflationary era explains why the universe, in its overall properties, looks the same in all directions: everything that we can see (and a good deal more than that) inflated from a single tiny region of space, converting its local properties into universal ones. Other advantages, which need not detain us here, accrue to the theory, at least among those who create model universes in their minds. One additional feature deserves emphasis, how-ever. The inflationary model makes a straightforward, testable prediction: space in the universe should be flat, neither positively nor negatively curved, but just as flat as our intuition imagines it.

According to this theory, the flatness of space arises from the enormous expansion that occurred during the inflation-ary epoch. Picture yourself, in analogy, on the surface of a balloon, and let the balloon expand by a factor so large that you lose track of the zeros. After this expansion, the part of the balloon's surface that you can see will be flat as a pan-cake. So, too, should all the space that we can ever hope to measure—if the inflationary model actually describes the real universe.

But the total density of matter—of all the dark matter plus all the ordinary matter—amounts to only about one-quarter of the amount required to make space flat. During the 1980s and 1990s, many theoretically minded cosmologists believed that because the inflationary model must be valid, new data would eventually close the cosmic "mass gap," the difference between the total density of matter, which pointed toward a negatively curved universe, and the critical density, seemingly required to achieve a cosmos with flat space. Their beliefs carried them buoyantly onward, even as observationally oriented cosmolo-gists mocked their overreliance on theoretical analysis. And then the mocking stopped.

In 1998, two rival teams of astrophysicists announced new observations implying the existence of a cosmological con-stant—not (of course) the same number that Einstein had pro-posed in order to keep the universe static, but another, quite different value, one that implies that the universe will expand forever at an ever-increasing rate.

If theorists had proposed this for yet another model universe, the world would have neither little noted nor long remembered their effort. Here, however, reputable experts in observing the real universe had mistrusted one another, checked on their rivals' suspicious activities, and discovered that they agreed on the data and their interpretation. The observational results not only implied a cosmological constant different from zero but also assigned to that constant a value that makes space flat.

What's that, you say? The cosmological constant flattens space? Aren't you suggesting, like the Red Queen in *Alice in Wonderland*, that we each believe six impossible things before breakfast? More mature reflection may, however, convince you that if apparently empty space does contain energy (!), that energy must contribute mass to the cosmos, just as Einstein's famous equation, $E = mc^2$, implies. If you've got some E, you can conceive it as a corresponding amount of m, equal to E divided by c^2. Then the total density must equal the total of the density contributed by matter, plus the density contributed by energy.

The new total density is what we must compare with the critical density. If the two are equal, space must be flat. This would satisfy the inflationary model's prediction of flat space, for space does not care whether its total density arises from the density of matter, or from the matter equivalent provided by the energy in empty space, or from a combination of the two.

———

The crucial evidence suggesting a non-zero cosmological constant, and thus the existence of dark energy, came from astrophysicists' observations of supernovae, stars that die spectacular deaths in titanic explosions. These supernovae, called either Type Ia or SN Ia, differ from other types, which occur when the cores of massive stars collapse after exhausting all possibilities of producing more energy by nuclear fusion. In contrast, SN Ia's owe their origin to stars that belong to binary star systems. Two stars that happen to be born close to each other will spend their lives performing simultaneous orbits around their common center of mass. If one of the two stars has more mass than the other, it will pass more rapidly through its prime of life, and in most cases will then lose its outer layers of gas, revealing its core to the cosmos as a shrunken, degenerate "white dwarf," an object no larger than Earth but containing as much mass as the Sun. Physicists call the matter in white dwarfs "degenerate" because it has such a high density—more than a hundred thousand times the density of iron or gold—that the effects of quantum mechanics act on matter in bulk form, preventing it from collapsing under its tremendous self-gravitational forces.

A white dwarf in mutual orbit with an aging companion star attracts gaseous material that escapes from the star. This matter, still relatively rich in hydrogen, accumulates on the white dwarf, growing steadily denser and hotter. Finally, when the temperature rises to 10 million degrees, the entire star ignites in nuclear fusion. The resulting explosion—similar in concept to a hydrogen bomb but trillions of times more violent—blows the white dwarf completely apart and produces a Type Ia supernova.

SN Ia's have proved particularly useful to astrophysicists by possessing two separate qualities. First, they produce the most luminous supernova explosions in the cosmos, visible across billions of light-years. Second, nature sets a limit to the maximum mass that any white dwarf can have, equal to about 1.4 times the Sun's mass. Matter can accumulate on a white dwarf's surface only until the white dwarf's mass reaches this limiting value. At that point, nuclear fusion blasts the white dwarf apart—and the blast occurs in objects with nearly the same mass and the same composition, strewn throughout the universe. As a result, all of these white-dwarf supernovae attain almost the same maximum energy output, and they all fade away at almost the same rate after they achieve their maximum brightness.

These dual attributes enable SN Ia's to provide astrophysicists with highly luminous, easily recognizable "standard candles," objects known to achieve the same maximum energy output wherever they appear. Of course, the distances to the supernovae affect their brightnesses as we observe them. Two SN Ia's, seen in two faraway galaxies, will appear to reach the same maximum brightness only if they have the same distance from us. If one has twice the distance of the other, it will attain only one-quarter of the other's maximum apparent brightness, because the brightness with which any object appears to us diminishes in proportion to the square of its distance.

Once astrophysicists learned how to use their detailed studies of the spectrum of light from Type Ia supernovae to recognize them, even billions of light-years away, they had a golden key to unlock the riddle of determining accurate distances. After measuring (through other means) the distances to the closest of the SN Ia's, they could estimate much greater distances to other Type Ia supernovae, simply by comparing the brightnesses of the relatively near and distant objects.

Throughout the 1990s, two teams of supernova specialists, one centered at Harvard and the other at the University of

California at Berkeley, refined this technique by finding how to compensate for the small but real differences among the SN Ia's, which the supernovae reveal to us through the details in their spectra. In order to use their newly forged key to unlock the distances to faraway supernovae, the researchers needed a telescope capable of observing distant galaxies with exquisite precision, and they found one in the Hubble Space Telescope, refurbished in 1993 in order to correct distorting effects from its primary mirror, which had been manufactured with the wrong shape. The supernova experts used ground-based telescopes to discover dozens of SN Ia's in galaxies billions of light-years from the Milky Way. They then arranged for the Hubble Telescope, for which they could obtain only a modest fraction of the total observing time, to study these newfound supernovae in detail.

As the 1990s drew toward a close, the two teams of supernova observers competed keenly to derive a new and expanded "Hubble diagram," the key graph in cosmology that plots the distances to galaxies versus the speeds at which the galaxies are moving away from us. Astrophysicists calculate these speeds through their knowledge of the Doppler effect, described below, which changes the colors of the galaxies' light by amounts that depend on the velocities at which the galaxies are receding from us.

Each galaxy's distance and recession velocity specify a point on the Hubble diagram. For relatively nearby galaxies these points march upward in a straight-line lockstep: a galaxy twice as distant from us as another turns out to be receding twice as fast. The direct proportionality between galaxies' distances and recession velocities finds algebraic expression in Hubble's law, the simple equation that describes the basic behavior of the universe: $v = H_0 \times d$. Here v stands for recession velocity, d for distance, and H is a universal constant, called the Hubble constant, that describes the entire universe at any particular time. (The subscript $_0$ denotes the present time.) Alien observers throughout the universe, studying the cosmos 14 billion years after the big bang, will find galaxies receding from them at speeds that follow Hubble's law, and all of them will derive the same value for the Hubble constant, though they will probably give it a different name. This assumption of cosmic democracy underlies all of modern cosmology. We cannot prove that the entire cosmos follows this democratic principle, but presuming a violation seems less reasonable than assuming cosmic uniformity. Far beyond our most distant horizon, the cosmos may behave quite differently from what we see. But cosmologists reject this approach, at least for the observable universe. In that case, $v = H_0 \times d$ represents universal law.

With time, however, the value of the Hubble constant can and does change. A new and improved Hubble diagram, one that extends to include galaxies many billions of light-years away, will provide not only the value of today's Hubble constant, H_0 (revealed by the slope of the line that runs through the points that represent galaxies' distances and recession velocities), but also the way in which the universe's current rate of expansion differs from its value billions of years ago. The latter value would be responsible for the details of the upper reaches of the graph, whose points describe the most distant galaxies ever observed. Thus a Hubble diagram extending to distances of many billion light-years would reveal the history of the expansion of the cosmos, embodied in its changing rate of expansion.

In striving for this goal, the astrophysics community struck a mother lode of good fortune in having two competing teams of supernova observers. The supernova results, first announced in February 1998, had an impact so great that no single group could have survived the natural skepticism of cosmologists to the overthrow of their widely accepted models of the universe. Because the two observing teams directed their skepticism primarily at each other, they brilliantly searched for errors in the other team's data or interpretation. When they pronounced themselves satisfied, despite their human prejudices, that their competitors were careful and competent, the cosmological world had little choice but to accept, albeit with some restraint, the news from the frontiers of space.

What was that news? Just that the most distant SN Ia's turned out a bit fainter than expected. This implies that the supernovae are somewhat farther away than they ought to be, which in turn shows that something made the universe expand a bit more rapidly than it should. What provoked this additional expansion? The only culprit that fits the facts is the "dark energy" that lurks in empty space—the energy whose existence corresponds to a non-zero value for the cosmological constant. By measuring the amount by which distant supernovae turned out to be fainter than expected, the two teams of astrophysicists measured the shape and fate of the universe.

———

When the two supernova teams achieved consensus, the cosmos turned out to be flat. To understand this bold conclusion, we must engage with cosmologists in their rough-and-tumble of Greek letters. To describe a universe with a non-zero cosmological constant requires one additional number. We must now add to the Hubble constant, which we write as H_0 to denote its value at the present time, and to the average density of matter, which by itself determines the curvature of space if the cosmological constant is zero, the density equivalent that the dark energy provides, which, by Einstein's formula $E = mc^2$, must possess the equivalent of mass (m) because it has energy (E). Cosmologists express the densities of matter and dark energy with the symbols Ω_M and Ω_Λ, where Ω (the Greek capital letter omega) stands for the ratio of the actual cosmic density to the critical density. The symbol Ω_M designates the ratio of the average density of all the matter in the universe to the critical density, while Ω_Λ stands for the ratio of the dark energy's density equivalent to the critical density. Here Λ (Greek capital lambda) represents the cosmological constant. In a flat universe, which has zero curvature of space, the sum of Ω_M and Ω_Λ always equals 1, because the total density—of actual matter plus the matter equivalent provided by the dark energy—exactly equals the critical density.

The observations of distant Type Ia supernovae measure the difference between Ω_M and Ω_Λ. Matter tends to slow the expansion of the universe, as gravity pulls everything toward everything else. The greater the density of matter, the more this mutual attraction will slow things down. Dark energy, however, does something quite different. Unlike pieces of matter, whose mutual attraction slows the cosmic expansion, dark energy has a strange property: it tends to make space expand, and thus to accelerate the expansion. As space expands, more dark energy comes into existence. In terms of dark energy, the expanding universe represents the ultimate free lunch. The new dark energy tends to make the cosmos expand still faster, so the free lunch grows ever larger as time goes on, and eventually swallows anything else that could affect the cosmic expansion. Since Ω_Λ measures the size of the cosmological constant, it provides us with the current magnitude of dark energy's expansionist ways.

When astrophysicists determined the relationship between galaxies' distances and their recession velocities, they found the result of the contest between gravity's pulling things together and dark energy's pushing them apart. Their measurements implied that $\Omega_\Lambda - \Omega_M = 0.46$, plus or minus about 0.03. Since astrophysicists had already determined that Ω_M equals approximately 0.25, this result sets Ω_Λ at about 0.71. Then the sum of Ω_Λ and Ω_M rises to 0.96, near the total predicted by the inflationary model. Later results refined these values and brought this sum even closer to the flat-universe value of 1. Today you would be hard-pressed to find a cosmologist who does not agree with the conclusion that the universe is flat.

Despite the agreement between the two competing groups of supernova experts, some cosmologists remained cautious. It is not every day that a scientist abandons a long-held belief, such as the conviction that the cosmological constant ought to be zero, and replaces it with a strikingly different one, such as the conclusion that dark energy fills every cubic centimeter of empty space. Almost all the skeptics who had followed the ins and outs of cosmological possibilities finally pronounced themselves convinced after they had digested new observations from a satellite designed and operated to observe the cosmic background radiation with unprecedented accuracy. That satellite, the highly significant WMAP described in Chapter 3, had accumulated sufficient data by 2003 for cosmologists to make a map of the entire sky, seen in the microwaves that carry most of the cosmic background radiation. Although earlier observations had revealed the basic results to be derived from this map, they had observed only small portions of the sky or had shown much less detail. WMAP's whole-sky map provided the capstone to the mapping effort, and firmly established the most important features of the cosmic background radiation.

The most striking and significant aspect of this map—also found by the balloon-borne observations of WMAP's predecessor, the COBE (COsmic Background Explorer) satellite, and by their successor, the Planck spacecraft—is its near-total featurelessness. No measurable differences in the intensity of the cosmic background radiation arriving from all different directions appear until we reach a precision of about one part in a thousand in our measurements. Even then, the only discernible differences appear as a slightly greater intensity, centered on one particular direction, that matches a corresponding slightly lesser intensity, centered on the opposite direction. These differences arise from the Doppler effect, the result of our Milky Way galaxy's motion relative to its neighbor galaxies. The Doppler effect causes us to receive slightly stronger radiation from the direction of this motion, not because the radiation actually *is* stronger, but because our motion toward the cosmic background radiation that we described in Chapter 3 slightly increases the energies of the photons that we detect.

Once we compensate for the Doppler effect, the cosmic background radiation appears perfectly smooth—until we attain an even higher precision of about one part in a hundred thousand. At that level, we observe tiny deviations from total smoothness. They track locations from which the CBR arrives with a bit more, or a bit less, intensity. As previously noted, the differences in intensity mark directions in which matter was either a little hotter and denser, or a little cooler and more rarefied, than the average value 380,000 years after the big bang. The COBE satellite first saw these differences; balloon-borne instruments and South Pole observations improved our measurements; and the WMAP and Planck satellites provided still better precision in surveying the entire sky. Their results have allowed cosmologists to construct a detailed map of the intensity of the cosmic background radiation, observed with unprecedented angular resolution of about 1 degree.

The tiny deviations from smoothness revealed by COBE, WMAP, and Planck have more than passing interest to cosmologists. First of all, they show the seeds of structure in the universe at the time when the cosmic background radiation ceased to interact with matter. The regions revealed as slightly denser than average at that time had a head start toward further contraction, and have won the competition to acquire the most matter by gravity. Thus the primary result from the new map of the CBR's intensity in different directions is the verification of cosmologists' theories of how the immense differences in density from place to place throughout the cosmos that we see now owe their existence to tiny differences in density that existed a few hundred thousand years after the big bang.

But cosmologists can use their new observations of the cosmic background radiation to discern another, still more basic fact about the cosmos. The details in the map of the CBR's intensity from place to place reveal the curvature of space itself. This amazing (how often cosmology calls forth this adjective!) result rests on the fact that the curvature of space affects how radiation travels through it. If, for example, space has a positive curvature, then when we observe the cosmic background radiation, our situation resembles that of an observer at the North Pole who looks along Earth's surface to study radiation produced near the Equator. Because the lines of longitude converge toward the pole, the source of radiation seems to span a smaller angle than it would if space were flat.

To understand how the curvature of space affects the angular size of features in the cosmic background radiation, imagine the time of decoupling, when the radiation finally ceased to interact with matter. At that time, the largest deviations from smoothness that could have existed in the universe had a size that cosmologists can calculate: the age of the universe at that time, multiplied by the speed of light—about 380,000 light-years across. This represents the maximum distance over which particles could have affected one another to produce any irregularities. At greater distances, the "news" from other particles would not yet have arrived, so they could not be responsible for any deviations from smoothness.

How large an angle would these maximum deviations span on the sky now? That depends on the curvature of space, which we can determine by finding the sum of Ω_M and Ω_Λ. The more closely that sum approaches 1, the more closely the space curvature will approach zero, and the larger will be the angular size that we observe for the maximum deviations from smoothness in the CBR. This space curvature depends only on the sum of the two Ωs, because both types of density make space curve in the same way. Observations of the cosmic background radiation therefore offer a direct measurement of $\Omega_M + \Omega_\Lambda$, in contrast to the supernova observations, which measure the *difference* between Ω_M and Ω_Λ.

The WMAP data showed that the largest deviations from smoothness in the CBR span an angle of about 1 degree, which implies that $\Omega_M + \Omega_\Lambda$ has a value of 1.02, plus or minus 0.02. Thus, within the limits of experimental accuracy, cosmologists found support for the conclusion that $\Omega_M + \Omega_\Lambda = 1$, so space is flat. The result from observations of distant SN Ia's may be stated as $\Omega_\Lambda - \Omega_M = 0.46$. Combination of this result with WMAP's conclusion that $\Omega_M + \Omega_\Lambda = 1$ implies that $\Omega_M = 0.27$ and $\Omega_\Lambda = 0.73$, with an uncertainty of a few percent in each number. Improved measurements from the Planck satellite's map of the CBR have refined these results, to $\Omega_M = 0.31$ and $\Omega_\Lambda = 0.69$. These numbers provide astrophysicists' current best estimates for the values of these two key cosmic parameters, whose uncertainty has now fallen to about plus or minus 2 percent. These parameters tell us that matter—both ordinary and dark—provides 31 percent of the total energy density in the universe, and dark energy 69 percent. (If we prefer to think of energy's mass equivalent, E/c^2, then dark energy furnishes 69 percent of all the mass.)

Cosmologists have long known that if the universe has a non-zero cosmological constant, the relative influence of matter and dark energy must change significantly as time passes and the amount of dark energy steadily increases. On the other hand, a flat universe remains flat forever, from its origin in the big bang to the infinite future that awaits us. In a flat universe, the sum of Ω_M and Ω_Λ always equals 1, so if one of these two changes, the other must also vary in compensation.

During the cosmic epochs that followed soon after the big bang, the dark energy produced hardly any effect on the universe. So little space existed then, in comparison to the eras that would follow, that Ω_Λ had a value just a bit above zero, while Ω_M was only a tiny bit less than 1. In those bygone ages, the universe behaved in much the same way as a cosmos without a cosmological constant. As time passed, however, Ω_M steadily decreased and Ω_Λ just as steadily increased, with their sum remaining constant at 1. Eventually, hundreds of billions of years from now, Ω_M will fall almost all the way to zero and Ω_Λ will rise nearly to unity. The history of flat space with a non-zero cosmological constant involves a continuous transition from its early years, when the dark energy barely mattered, through the "present" period, when Ω_M and Ω_Λ have roughly equal values, and on into an infinitely long future, when matter will spread so diffusely through space that Ω_M must pursue an infinitely long slide toward zero, even as the sum of the two Ωs remains equal to 1.

Observational deduction of how much mass exists in galaxy clusters now gives Ω_M a value of about 0.29, while the observations of the CBR and distant supernovae yield a value close to 0.31. Within the limits of experimental accuracy, these two values coincide. If the universe in which we live does have a non-zero cosmological constant, and if that constant is responsible (along with the matter) for producing the flat universe that the inflationary model predicts, then the cosmological constant must have a value that makes Ω_Λ close to 0.7, more than twice the value of Ω_M. In other words, Ω_Λ must now do most of the work in making $\Omega_M + \Omega_\Lambda$ equal to 1. This means that we have already passed through the cosmic era when matter and the cosmological constant contributed the same amount (with each of them equal to 0.5) toward maintaining the flatness of space.

In less than a decade of astronomical investigation, the double-barreled blast from the Type Ia supernovae and the cosmic background radiation changed the status of dark energy from a far-out idea that Einstein once toyed with to a cosmic fact of life. Unless a mass of observations eventually prove to be misinterpreted, inaccurate, or just plain wrong, we must accept the result that the universe will never contract or recycle itself. Instead, the future seems bleak: a hundred billion years from now, when most stars will have burnt themselves out, all but the closest galaxies will have vanished across our horizon of visibility.

By then, the Milky Way will have coalesced with its nearest neighbors, creating one giant galaxy in the literal middle of nowhere. Our night sky will contain orbiting stars, dead and alive, and nothing else, leaving future astrophysicists a cruel universe. With no galaxies to track the cosmic expansion, they will erroneously conclude, as did Einstein, that we live in a static universe. The cosmological constant and its dark energy will have evolved the universe to a point where they cannot be measured or even dreamt of.

Enjoy cosmology while you can.

6

TENSION IN THE COSMOS!

During the years that followed the discovery of dark energy, you could forgive cosmologists for placing dark energy close to, or even at, the top of their list of vexing issues. This period has provided what astrophysicists always welcome: increasingly accurate measurements of the parameters that describe the cosmos, from its smallest to its largest size scales, from times soon after its origin to the present situation. In their efforts, cosmologists have continued to seek greater precision in a basic cosmic parameter, the universe's rate of expansion. They have developed two distinct methods, almost equal in accuracy, to measure this rate. Their labors, however, have apparently produced another troubling, provocative, and potentially fruitful issue, because the two methods have yielded divergent outcomes.

Conflicting results such as these typically awaken two mutually contradictory reactions, sometimes within the same person. On the one hand, the divergence may well have arisen not from the cosmos but instead from misinterpretation, errors, or overestimates of the precision of the data obtained through imaginative hard work. On the other, if this particular divergence proves real, it raises the possibility of a new understanding of the universe, either through improved knowledge of its history or—even more exciting—through a change in our comprehension of the basic physics that underpins every cosmological analysis of astrophysicists' observations.

The divergence in question concerns the value of the most basic parameter of modern cosmology: the Hubble constant, H_0, that expresses the rate at which the universe now expands. With some risk of over-nerdiness, we may note that astrophysicists describe the Hubble constant in units of "kilometers per second per megaparsec." These units enumerate the amount by which galaxies' recession velocities, measured in kilometers per second, increase with their increasing distances, measured in megaparsecs, each of which spans 3.26 million light-years. One method for determining the Hubble constant now provides a value a bit more than 67 kilometers per second per megaparsec, but its chief competitor yields another, about 10 percent greater, close to 73. The difference between these two numbers has led to a situation that cosmological aficionados often describe as "cosmic tension." "Crisis in cosmology" would garner still greater attention, but for now, we may rest content with "tension" as we ask: What does this mean for us, and for science?

Taking a long view of history, we might be tempted to celebrate astrophysicists' convergence toward a single, agreed-upon value rather than worrying about a troubling divergence. Before the Hubble Telescope, prominent astrophysicists who determined the Hubble constant's value disagreed by a factor of two, with one side favoring a value of 50 and the other 100. The fact that today the difference between 67 and 73 troubles many cosmologists provides glorious testimony to how far we have come in a single lifetime.

Some astrophysicists—typically not among those not directly involved in the measurements or their interpretation—currently radiate this calm view of the situation. They judge it likely that before long, a rather prosaic resolution of the cosmic tension will lead to a local cosmic relaxation, with—could it be?—a value close to 70 emerging as the correct one. But many of those who have spent years and even decades striving to find the precise value of the Hubble constant take an opposing view (as we might well have predicted), and continue to engage in a cosmic struggle over the divergence. If both camps prove correct in their numbers, then from history's vantage point, the present may well come to be seen as an epoch in which two numbers that could not be reconciled opened the door to new physics.

What are the two observational approaches that have given birth to the cosmic tension? The first of these, which employed distance estimates obtained from observations of supernova explosions in faraway galaxies, revealed the existence of dark energy. Ever-better observations of these exploding stars, together with refinements in astrophysicists' understanding of the subtle differences among them, have led to a convergence on the value close to 73. Before we discuss the uncertainties that surround this value, however, we should examine the chief alternative method for determining the Hubble constant.

This approach relies on the use of what cosmologists call a "standard ruler," an analogy to the "standard candles" embodied in the supernovae used for the conventional approach to the Hubble constant. As described in the previous chapter, at the time of decoupling, 380,000 years after the big bang, the homogenizing effects that radiation exerted on matter essentially ceased. From then on, radiation roamed freely among particles of matter without affecting them significantly. This occurred when the maximum distance within which particles of matter could affect one another spanned about 420,000 light-years, because regions more widely separated than this had not had time to communicate in any way. This distance gives cosmologists their standard ruler. We noted its existence in the previous chapter as the maximum size of the deviations from smoothness in the cosmic background radiation.

As space expanded, so did the standard ruler, which continued to measure the largest spans of space within which coherent deviations of the density of matter from its average value could arise. We can now "see" the ruler—more precisely, its effects—in two different epochs. We have already met the first of these: tiny departures from uniformity in the CBR that track the slightly uneven distribution of matter at the time of decoupling. During the following billions of years, these one-part-in-a-hundred-thousand deviations in density evolved to become immensely larger differences between the densities of matter within giant clusters of galaxies and in the regions between them. The maximum sizes of these clusters show the amount by which the size of the standard ruler has increased from the time of decoupling to the present.

The second method for determining the Hubble constant therefore aims to create an accurate map of the universe today for comparison with the original differences in the cosmic background radiation. (In actual fact, "today" means "only a couple of billion years ago," the average look-back time to the clusters of galaxies that have grown from the tiny deviations embodied in the CBR.) During the first decades of the twenty-first century, in an effort that continues to gain precision, a project named the Sloan Digital Sky Survey has employed a dedicated telescope at Apache Point, New Mexico, to map the three-dimensional distribution of galaxies in space with unprecedented accuracy, and thus to provide the modern size of the standard ruler, which turns out to be approximately 490 *million* light-years. Comparison of this distance with the ruler's 450 *thousand* light-years at the time of decoupling leads to a value of the Hubble constant close to 67.

How much uncertainty can we assign to each of these two clashing numbers, 67 and 73? The most recent analysis by the teams of astrophysicists who use the standard-ruler approach produces a value of 67.3, plus or minus 0.6. The alternative, supernova-based approach to determining the Hubble constant involves several independent teams of supernova observers, competing not only for the most accurate results but also for the most appealing team acronym, two of which, HOLiCOW and SH0ES, sweetly work H_0 into their names. The most recent SH0ES value comes in as 73.3, plus or minus 1.0, whereas the HOLiCOW effort yields 73.3, plus or minus 1.8. The difference between 67-plus and 73-plus, along with their estimated uncertainties, creates what scientists call a "five-sigma difference," which we may translate as "too large to be easily resolved." (Most scientists regard a three-sigma divergence as highly significant, provided that they trust the underlying data.)

———

Before we confront these opposing outcomes, we should note, perhaps with some surprise, that astrophysicists have three additional approaches in their observational arsenal to determine the value of the Hubble constant. One of these has already been deployed, while two others should soon help to improve our knowledge.

The more developed effort involves a new way to estimate the distances to supernovae in comparatively nearby galaxies, by making detailed observations of the brightest stars in mammoth star clusters within the closer galaxies that contain supernovae. From their understanding of how stars evolve, astrophysicists know the energy output of these stars. As with observations of supernovae in galaxies, a comparison of the observed brightnesses of objects known to have the same intrinsic brightness yields the ratios of the objects' distances. Although this observational program does not extend to distances as great as those attainable with supernova observations, its results suggest that the Hubble constant has the compromise value of 70 mentioned above. An analysis of these results by the proponents of the 73 value concludes that in fact they only slightly reduce that number. From such conflicts, eventual resolution may arise.

Meanwhile, two additional, independent methods to calibrate the Hubble constant have achieved some success, but for now rest in comparative infancy. Both methods break new ground by relying on Einstein's general theory of relativity. One involves the bending of space by gravitational forces, while the other looks to the gravitational radiation that scientists have detected only within the past few years. Like the older and better-established techniques, both methods aim to determine more accurate distances to objects for comparison with the speeds with which they are receding from us. The first of these analyzes data on the slight gravitational bending of the cosmic background radiation in passing by a multitude of galaxies on its way to us. The second, described more fully in Chapter 9, deals with observations of "standard sirens," a subset of sources of gravitational radiation with similar characteristics throughout the observed cosmos. Standard sirens are named in analogy with the cosmological "standard rulers" that we have already encountered. Both the space-bending and standard-siren approaches promise results that, before long, should rival those from the currently best techniques for measuring the universe's rate of expansion.

———————

Taking the broad view, how should we judge the significance of the current tension in cosmology? Like astrophysicists themselves, astute readers may reasonably search their own biases in predicting a possible resolution. Do you favor a conservative approach, remain calm, and expect that all values will converge to 70 before long? Or do you favor a revolution: a provable confrontation between the values of 67 and 73 that will open the door to new physics? We can be sure that the cosmos itself has no crisis. The problems arise on Earth, where human understanding inevitably falls well short of perfection. Cosmologists and physicists who see the tension as crying out for solution have attempted, as their job descriptions require, to resolve it by determining what we have missed in our understanding of the universe.

Perhaps to their creators' credit, the list of proposed solutions would overtax most readers. Almost all such suggestions either change the currently accepted model of the universe's expansion history or introduce "new physics," which could include changing relativity theory or the laws of gravitation. The most popular new-physics suggestions involve hypothetical, unknown particles (entirely different from the hypothetical, unknown particles that form the dark matter) or hypothetical, subtle changes in the amount of dark energy during the early expansion of the universe, either before the time of decoupling or soon thereafter. Unfortunately for some of these theories, though fortunately for the progress of science, the precision of our current observations of the cosmic background radiation places strong limits on these hypotheses, and in the most straightforward cases, eliminates them with a high degree of probability. From a certain perspective, this increases the excitement that the cosmic tension brings to cosmology: we may find not only that new physics lurks within the seemingly modest disagreement between 67 and 73, but also that the addition of "simple" new physics may prove inadequate. In that case, a wider revision of our understanding would have to occur in order for the cosmic tension to resolve itself, allowing astrophysicists to concentrate on the new conundrums that will doubtless arise from future observations.

7

ONE UNIVERSE OR MANY?

We may obtain some relief from the headache of cosmo-logical tension by reflecting that whatever the value of H_0 may be, we should remain confident in two cosmic facts: we live in an expanding universe, and this expansion is accelerating. Refutation of either of these cosmic characteristics would usher in a cosmic revolution more profound than the acceptance of any new physics. To salute the discovery that we live in an accelerating universe, and to understand why astrophysicists have such trust in this conclusion, we should look back to the time when the cosmic acceleration rocked the world of cosmology, and perceive why the accelerating universe quickly gained widespread acceptance.

The news broke early in 1998, with the first announcement of the supernova observations that point to this acceleration. After the accelerating universe received confirmation from increasingly detailed observations of the cosmic background radiation, and now that cosmologists have had several decades to wrestle with the implications of an accelerating cosmic expansion, two great questions have emerged to bedevil their days and brighten their dreams: What makes the universe accelerate? And why does that acceleration have the particular value that now characterizes the cosmos?

The simple answer to the first question assigns all responsibility for the acceleration to the existence of dark energy or, equivalently, to a non-zero cosmological constant. The amount of acceleration depends directly on the amount of dark energy per cubic centimeter: more energy implies greater acceleration. Thus, if cosmologists could only explain where the dark energy comes from, and why it exists in the amount that they find today, they could claim to have uncovered a fundamental secret of the universe—the explanation for the cosmic "free lunch," the energy in empty space that continuously drives the cosmos toward an eternal, ever-more-rapid expansion and a far future of enormous amounts of space, correspondingly enormous amounts of dark energy, and almost no matter per cubic light-year.

What makes dark energy? From the deep realms of particle physics, cosmologists can produce an answer: if we trust what we have learned from the quantum theory of matter and energy, dark energy arises from events that must occur in empty space. All of particle physics rests on this theory, which has been verified so often and so exactly in the submicroscopic realm that almost all physicists accept it as correct. An integral part of quantum theory implies that what we call empty space in fact buzzes with "virtual particles," which wink into and out of existence so rapidly that we can never pin them down directly, though we can observe their effects. The continual appearance and disappearance of these virtual particles, called the "quantum fluctuations of the vacuum" by those who like a fine physics phrase, gives energy to empty space. Furthermore, particle physicists can, without much difficulty, calculate the amount of energy that resides in every cubic centimeter of the vacuum. The straightforward application of quantum theory to what we call a vacuum predicts that quantum fluctuations must create dark energy. When we tell the story from this perspective, the great question about dark energy seems to be, Why did cosmologists take so long to recognize that this energy must exist?

Unfortunately for attempts to find a simple explanation of dark energy, the details of the actual situation turn this question into, How did particle physicists go so far wrong? Calculations of the amount of dark energy that lurks in every cubic centimeter produce a value about 120 powers of ten greater than the value that cosmologists have found from observations of supernovae and the cosmic background radiation. In far-out astronomical situations, calculations that prove correct to within a single factor of 10 are often judged at least temporarily acceptable, but a factor of 10^{120} cannot be swept under the rug, even by physics Pollyannas. If real empty space contained dark energy in anything like the amounts proposed by particle physics, the universe would have long since puffed itself into so large a volume that our heads could never have begun to spin, since a tiny fraction of a second would have sufficed to spread matter out to unimaginable rarefaction. Theory and observation agree that empty space ought to contain dark energy, but they disagree by a trillion to the 10th power about the amount of that energy. No earthly analogy, or even a cosmic one, can illustrate this discrepancy accurately. The distance to the farthest galaxy that we know exceeds the size of a proton by a factor of 10^{40}. Even this enormous number is only the cube root of the factor by which theory and observation currently diverge concerning the value of the cosmological constant.

Particle physicists and cosmologists have long known that quantum theory predicts an unacceptably large value for the dark energy, but in the days when the cosmological constant was thought to be zero, they hoped to discover some explanation that would, in effect, cancel positive with negative terms in the theory and thereby finesse the problem out of existence. A similar cancellation once solved the problem of how much energy virtual particles contribute to the particles that we do observe. Now that the cosmological constant turns out to be non-zero, the hopes of finding such a cancellation seem dimmer. If the cancellation does exist, it must somehow remove almost all of the mammoth theoretical value we have today. For now, lacking any good explanation for the size of the cosmological constant, cosmologists must continue to collaborate with particle physicists as they seek to reconcile theories of how the cosmos generates dark energy with the value that astrophysicists have obtained for the amount of dark energy per cubic centimeter.

Since the turn of the century, some of the finest minds engaged in cosmology and particle physics have directed much of their energy toward explaining this observational value, with no success at all. This provokes fire, and sometimes ire, among theorists, in part because they know that a Nobel Prize—not to mention the immense joy of discovery—awaits those who can explain what nature has done to make space as we find it. But yet another issue stokes intense controversy as it cries out for explanation: Why does the amount of dark energy, as measured by its mass equivalent, roughly equal the amount of energy provided by all the matter in the universe?

We can recast this question in terms of the two Ωs that we use to measure the density of matter and the density equivalent of dark energy: Why do Ω_M and Ω_Λ roughly equal each other, rather than one being enormously larger than the other? During the first billion years after the big bang, Ω_M was almost precisely equal to 1, while Ω_Λ was essentially zero. In those years, Ω_M was first millions, then thousands, and afterward hundreds of times greater than Ω_Λ. Today, with $\Omega_M = 0.31$ and $\Omega_\Lambda = 0.69$, the two values are roughly equal, though Ω_Λ is already notably larger than Ω_M. In the far future, more than 50 billion years from now, Ω_Λ will become first hundreds, then thousands, after that millions, and still later billions of times greater than Ω_M. Only during the cosmic era from about 3 billion to 50 billion years after the big bang do the two quantities even approximately match each other.

To the easygoing mind, the interval between 3 billion and 50 billion years embraces quite a long period of time. So what's the problem? From an astronomical viewpoint, this stretch of time amounts to nearly nothing. Astrophysicists often take a logarithmic approach to time, dividing it into intervals that steadily increase by factors of 10. First the cosmos had some age; then it grew 10 times older; then 10 times older than that; and so on toward infinite time, which requires an infinite number of 10-times jumps. Suppose that we start counting time at the earliest moment after the big bang that has any significance in quantum theory, 10^{-43} second after the big bang. Since each year contains about 30 million (3×10^7) seconds, we need about 60 factors of 10 to pass from 10^{-43} second to 3 billion years after the big bang. In contrast, we require only a bit more than a single factor of 10 to stroll from 3 billion to 50 billion years, the only period when Ω_M and Ω_Λ are roughly equal. After that, an infinite number of 10-times factors opens the way to the infinite future. From this logarithmic perspective, only a vanishingly small probability exists that we should find ourselves living in a cosmic situation for which Ω_M and Ω_Λ have even vaguely similar values. Michael Turner, a leading American cosmologist, has termed this conundrum—the question of why we find ourselves alive at a time when Ω_M and Ω_Λ are approximately equal—the "Nancy Kerrigan problem" in honor of the Olympic figure skater who asked, after enduring an assault by her rival's boyfriend, "Why me? Why now?"

Despite their inability to calculate a cosmological constant whose value comes anywhere close to the measured one, cosmologists do have an answer to the Kerrigan problem, but they differ sharply on its significance and implications. Some embrace it; some accept it only reluctantly; some dance around it; and some despise it. This explanation links the existence of life on Earth to the universal value of the cosmological constant: because we are here, alive on a planet that orbits an average star in an average galaxy, the parameters that describe the cosmos, and in particular the value of the cosmological constant, must have values that allow us to exist.

Consider, for example, what would happen in a universe with a cosmological constant much larger than its actual value. A much greater amount of dark energy would make Ω_Λ rise far above Ω_M, not after about 50 billion years but after only a few million years. By this time, in a cosmos dominated by the accelerating effects of dark energy, matter would spread so rapidly apart that no galaxies, stars, or planets could form. If we assume that the stretch of time from the first formation of clumps of matter to the origin and development of life covers at least 1 billion years, we can conclude that our existence limits the cosmological constant to a value between zero and a few times its actual value, while ruling out of play the infinite range of higher values.

This argument gains more traction if we assume, as do many cosmologists, that everything we call the universe belongs to a much larger "multiverse," which quite possibly contains an infinite number of universes, none of which interact with any other: in the multiverse concept, the entire state of affairs embeds in higher dimensions, so space in our universe remains completely inaccessible to any other universe, and vice versa. This lack of even theoretically possible interactions puts the multiverse theory into the category of apparently nontestable, and therefore nonverifiable, hypotheses—at least until wiser minds find ways to test the multiverse model. In the multiverse, new universes are born at completely random times, capable of swelling up by inflation into enormous volumes of space, and of doing so without interfering in the least with the infinite number of other universes.

In the multiverse, each new universe springs into existence with its own physical laws and its own set of cosmic parameters, including the rules that determine the size of the cosmological constant. Many of these other universes have cosmological constants enormously larger than ours, and quickly accelerate themselves into situations of near-zero density, no good for life. Only a tiny, perhaps an infinitesimal, fraction of all the universes in the multiverse furnish conditions that allow life to exist, because only this fraction have parameters that permit matter to organize itself into galaxies, stars, and planets, and for those objects to last for billions of years.

Cosmologists call this approach to explaining the value of the cosmological constant the anthropic principle, though the "anthropic approach" probably offers a better name. This approach toward explaining a crucial question in cosmology has one great virtue: people love it or hate it, but rarely feel neutral

about it. Like many intriguing ideas, the anthropic approach
can be bent to favor, or at least seem to favor, various theological and teleological mental edifices. Some religious fun-
damentalists find that the anthropic approach supports their
beliefs because it implies a central role for humanity: without
someone to observe it, the cosmos—at least the cosmos that we
know—would not, could not, be here; hence a higher power must
have made things just right for us. An opponent of this conclusion would note this is not really what the anthropic approach
implies; on a theological level, this argument for the existence
of God implies surely the most wasteful creator one might imagine, who makes countless universes in order that in a tiny sector
of just one of these, life might arise. Why not skip the middle-
man and follow older creation myths that center on humanity?

On the other hand, if you choose to see God in everything,
as Spinoza did, you cannot help but admire a multiverse that
effloresces with universes without end. Like most news from
the frontier of science, the concept of a multiverse, and the
anthropic approach, can be easily bent in different directions
to serve the needs of particular belief systems. As things
stand, many cosmologists find the multiverse quite enough to
swallow without connecting it to any system of beliefs. Stephen Hawking, who (like Isaac Newton before him) held the
Lucasian chair in astronomy at Cambridge University, judged
the anthropic approach an excellent resolution of the Kerrigan
problem. The late Steven Weinberg, who won the Nobel Prize
for his insights into particle physics, disliked this approach but
pronounced himself in favor, because and so long as no other
reasonable solution has appeared.

History may eventually show that for now, cosmologists are
concentrating on the wrong problem—wrong in the sense that

we don't yet understand enough to attack it properly. Weinberg liked the comparison with Johannes Kepler's attempt to explain why the Sun has six planets (as astrophysicists then believed) and why they move in the orbits that they do. Four hundred years after Kepler, astrophysicists still know far too little about the origin of planets to explain the precise number and orbits of the Sun's family. We do know that Kepler's hypothesis, which proposed that the spacing of the planets' orbits around the Sun allows one of the five perfect solids to fit exactly between each pair of adjoining orbits, has no validity whatsoever, because the solids do not fit particularly well, and (even more important) because we have no good reason to explain why the planets' orbits should obey such a rule. Later generations may regard today's cosmologists as latter-day Keplers, struggling valiantly to explain what remains inexplicable by today's understanding of the universe.

Not everyone favors the anthropic approach. Some cosmologists attack it as defeatist, ahistorical (since this approach contradicts numerous examples of the success of physics in explaining, sooner or later, a multitude of once-mysterious phenomena), and dangerous, because the anthropic approach smacks of intelligent-design arguments. Furthermore, many cosmologists find unacceptable, as grounds for a theory of the universe, the assumption that we live in a multiverse that contains a multitude of universes with which we can never interact, even in theory.

The debate over the anthropic principle highlights the skepticism that underlies the scientific approach to understanding the cosmos. A theory that appeals to one scientist, typically the one who thought it up, may seem ridiculous, or just plain wrong, to another. Both know that theories survive and thrive when other scientists find them best at explaining most of the observational data. As a famous scientist once remarked, Beware of a theory that explains *all* the data, because some of the data will quite likely turn out to be wrong.

The future may not produce a quick resolution to this debate, but it will surely bring forth other attempts to explain what we see in the universe. For example, Paul Steinhardt of Princeton University, who could use some tutoring in creating catchy names, has produced a theoretical "ekpyrotic model" of the cosmos in collaboration with Neil Turok of the University of Edinburgh. Motivated by the section of particle physics called string theory, Steinhardt envisions a universe with 11 dimensions, most of which are "compactified," more or less rolled up like a sock, so that they occupy only infinitesimal amounts of space. But some of the additional dimensions have real size and significance, except that we can't perceive them because we remain locked into our familiar four. If you pretend that all of space in our universe fills an infinite thin sheet (this model reduces the three dimensions of space to two), you can imagine another, parallel sheet, and then picture the two sheets approaching and colliding. The collision produces the big bang, and as the sheets rebound from each other, each sheet's history proceeds along familiar lines, giving birth to galaxies and stars. Eventually, the two sheets cease to separate and start to approach each other again, producing another collision and another big bang in each sheet. The universe thus has a cyclical history, repeating itself, at least in its broadest outlines, at intervals of hundreds of billions of years. Since *ekpyrosis* means "conflagration" in Greek (recall the more familiar word "pyromaniac"), the "ekpyrotic universe" reminds all those with Greek at the tips of their brains of the great fire that gave birth to the cosmos that we know.

This ekpyrotic model of the universe has emotional and intellectual appeal, though not enough to win the hearts and minds of many of Steinhardt's fellow cosmologists. Not yet, anyhow. Something vaguely like the ekpyrotic model, if not this model itself, may someday offer the breakthrough that cosmologists now await in their attempts to explain the dark energy. Even those who favor the anthropic approach would hardly dig in their heels to resist a new theory that could provide a good explanation for the cosmological constant without invoking an infinite number of universes, of which ours happens to be one of the lucky ones.

Attempts to explain the cosmological constant, or any other physical parameter that describes the universe, remind us that each new detection of an aspect of the cosmos provokes new questions that demand further answers, some of which will require many years to resolve. Cynics may ask, who cares? Not only the anthropic principle but the entire sweep of cosmology as well, from the Greeks to the multiverse, has had almost no impact on our daily lives. Nevertheless, cosmological questions continue to fascinate many of us, bringing us face to face (as F. Scott Fitzgerald wrote at the end of *The Great Gatsby*) with something commensurate to our capacity to wonder.

For those who seek knowledge closer to home, it's time to turn our attention to individual cosmic objects, some of which indeed have an immediate impact on our lives. We may start, as the universe did, with the formation of the largest entities in the universe before proceeding toward the local, cosmically insignificant ones that we can explore on firm ground.

PART II
THE ORIGIN OF GALAXIES AND COSMIC STRUCTURE

8

DISCOVERING GALAXIES

Two and a half centuries ago, shortly before the English astronomer Sir William Herschel built the world's first seriously large telescope, the known universe consisted of little more than the stars, the Sun and Moon, the planets, a few moons of Jupiter and Saturn, some fuzzy objects, and the galaxy that forms a milky band across the night sky. Indeed, the word "galaxy" derives from the Greek *galaktos*, or "milk." The sky also held the fuzzy objects, scientifically named nebulae after the Latin word for clouds—objects of indeterminate shape such as the Crab nebula in the constellation Taurus, and the Andromeda nebula, which appears to live among the stars of the constellation Andromeda.

Herschel's telescope had a mirror 48 inches across, an unprecedented size for 1789, its year of completion. A complex array of trusses to support and point this telescope made it an ungainly instrument, but when he aimed it at the heavens, Herschel could readily see the countless stars that compose the Milky Way. Using his 48-incher, as well as a smaller, nimbler telescope, Herschel and his sister Caroline compiled the first extensive "deep sky" catalogue of northern nebulae. Sir

John—Herschel's son—continued this family tradition, adding to his father and aunt's list of northern objects and, during an extended stay at the Cape of Good Hope at the southern tip of Africa, cataloguing some 1,700 fuzzy objects visible from the Southern Hemisphere. In 1864, Sir John produced a synthesis of the known deep-sky objects: *A General Catalogue of Nebulae and Clusters of Stars*, which included more than 5,000 entries. In spite of that large body of data, nobody at the time knew the true identity of the nebulae, their distances from Earth, or the differences among them. Nevertheless, the 1864 catalogue made it possible to classify the nebulae morphologically—that is, according to their shapes. In the "we call 'em as we see 'em" tradition of baseball umpires (who came into their own at just about the time that Herschel's *General Catalogue* was published), astronomers named the spiral-shaped nebulae "spiral nebulae," those with a vaguely elliptical shape "elliptical nebulae," and the various irregularly shaped nebulae—neither spiral nor elliptical—"irregular nebulae." Finally, they called the nebulae that looked small and round, like a telescopic image of a planet, "planetary nebulae," a mistake in nomenclature that has permanently confused newcomers to astronomy.

For most of its history, astronomy remained plainspoken, using descriptive methods of inquiry that greatly resembled those used in botany. Using their lengthening compendia of stars and fuzzy things, astronomers searched for patterns and sorted objects according to them. Quite a sensible step, too. Most people, beginning in childhood, arrange things according to appearance and shape without even being told to do so. But this approach can carry you only so far. The Herschels always assumed that because many of their fuzzy objects span about the same size on the night sky, all the nebulae must lie at about the same distance from Earth. So to them it was simply good, evenhanded science to subject all the nebulae to the same rules of sorting.

Trouble is, the assumption that all nebulae lay at similar distances turned out to be badly mistaken. Nature can be elusive, even devious. Some of the nebulae classified by the Herschels are no farther away than the stars, so they are relatively small (if a few trillion miles across can be called "relatively small"). Others have turned out to be much more distant, so they must be much larger than the fuzzy objects relatively close to us if they are to appear the same size on the sky.

The take-home lesson is that at some point you've got to stop fixating on what something looks like and start asking what it is. Fortunately, by the late nineteenth century, advances in science and technology had empowered astronomers to do just that, to move beyond merely classifying the contents of the universe. That shift led to the birth of astrophysics, the useful application of the laws of physics to astronomical situations.

During the same era when Sir John Herschel published his vast catalogue of nebulae, a new scientific instrument, the spectroscope, joined the search for nebulae. The sole job of a spectroscope is to break light into a rainbow of its component colors. Those colors, and features embedded within them, reveal not only fine details about the chemical composition of the light source but also, because of the Doppler effect described in Chapter 5, the motion of the light source toward or away from Earth.

Spectroscopy eventually revealed something remarkable: the spiral nebulae, which happen to predominate outside the swath of the Milky Way, are nearly all moving away from Earth, and at extremely high speeds. In contrast, all the planetary nebulae, as well as most irregular nebulae, are traveling at relatively low speeds—some toward us and some away from us. Had some catastrophic explosion taken place in the center of the Milky Way, booting out only the spiral nebulae? If so, why weren't any of them falling back? Were we catching the catastrophe at a special time? In spite of advances in photography that brought forth faster emulsions, enabling astrophysicists to measure the spectra of ever-dimmer nebulae, the exodus continued and these questions remained unanswered.

Most advances in astronomy, as in other sciences, have been driven by the introduction of better technology. As the 1920s opened, another key instrument appeared on the scene: the formidable 100-inch Hooker Telescope at the Mount Wilson Observatory near Pasadena, California. In 1923, the American astrophysicist Edwin P. Hubble used this telescope—the largest in the world at that time—to find a special breed of star, a Cepheid variable star, in the Andromeda nebula. Variable stars of any type vary in brightness according to well-known patterns; Cepheid variables, named for the prototype of the class, a star in the constellation Cepheus, are all extremely luminous and therefore visible over vast distances. Because their brightness varies in recognizable cycles, patience and persistence will reveal increasing numbers of them to the careful observer. Hubble had found a few of these Cepheid variables within the Milky Way and estimated their distances; yet, to his astonishment, the Cepheid he found in Andromeda was much dimmer than any of those.

The most likely explanation for this dimness was that the new Cepheid variable, and the Andromeda nebula in which it lives, sits at a distance much greater than those to the Cepheids in the Milky Way. Hubble realized that this placed the Andromeda nebula at so great a distance that it could not possibly lie among the stars in the constellation Andromeda, nor anywhere within the Milky Way—and could not have been kicked out, along with all its spiral sisters, during a catastrophic milk spill.

The implications were breathtaking. Hubble's discovery showed that spiral nebulae are entire systems of stars in their own right, as huge and as packed with stars as our own Milky Way. In the phrase of the philosopher Immanuel Kant, Hubble had demonstrated that "island universes" by the dozens must lie outside our own star system, for the object in Andromeda merely led the list of well-known spiral nebulae. The Andromeda nebula was, in fact, the Andromeda *galaxy*.

———

By 1936, enough island universes had been identified and photographed through the Hooker and other large telescopes that Hubble, too, decided to try his hand at morphology. His analysis of galaxy types rested upon the untested assumption that variations from one shape to another among galaxies signify evolutionary steps from birth to death. In his 1936 book, *Realm of the Nebulae*, Hubble classified galaxies by placing the different types along a diagram shaped like a musical tuning fork, whose handle represents the elliptical galaxies, with rounded ellipticals at the far end of the handle and flattened ellipticals near the point where the two tines join. Along one tine lie the ordinary spiral galaxies: those nearest the handle have their spiral arms wound extremely tightly, while those toward the tine's end have increasingly loosely wound spiral arms. Along the other tine are spiral galaxies whose central region displays a straight "bar" but are otherwise similar to ordinary spirals.

Hubble imagined that galaxies start their lives as round ellipticals and become flatter and flatter as they continue to take shape, ultimately revealing a spiral structure that slowly unfurls with the passage of time. Brilliant. Beautiful. Even elegant. But just plain wrong. Not only were entire classes of irregular galaxies omitted from this scheme, but astrophysicists would later learn that the oldest stars in almost every galaxy were about the same age, implying that all the galaxies were born during a single era in the history of the universe.

For three decades (with some research opportunities lost because of World War II), astrophysicists observed and catalogued galaxies in accordance with Hubble's tuning-fork diagram as ellipticals, spirals, and barred spirals, with irregulars a minority subset, completely off the chart because of their strange shapes. Of elliptical galaxies one might say, as Ronald Reagan did about California's redwoods, that when you've seen one, you've seen them all. Elliptical galaxies resemble one another in possessing neither the spiral-arm patterns that characterize spirals and barred spirals, nor the giant clouds of interstellar gas and dust that give birth to new stars. In these galaxies, star formation ended many billion years ago, leaving behind spherical or ellipsoidal groups of stars. The largest elliptical galaxies, like the largest spirals, each contain many hundred billion stars—perhaps even a trillion or more—and have diameters close to a hundred thousand light-years. With the exception of professional astrophysicists, no one has ever sighed over the fantastic patterns and complex star-formation histories of elliptical galaxies for the excellent reason that, at least in comparison with spirals, ellipticals have simple shapes and straightforward star formation: they all turned gas and dust into stars until they could do so no more.

Fortunately, spirals and barred spirals provide the visual excitement so lacking in ellipticals. The most deeply resonant of all the galaxy images that we may ever see, a view of the entire Milky Way taken from outside it, will stir our hearts and minds, just as soon as we manage to send a camera to distances several hundred thousand light-years above or below the central plane of our galaxy. Today, when our most far-flung space probes have traveled a billionth of that distance, this goal may seem unattainable, and indeed even a probe that could reach nearly the speed of light would require a long wait—far longer than the current span of recorded history—to yield the desired result. For the time being, astrophysicists must continue to map the Milky Way from inside, sketching the galactic forest by delineating its stellar and nebular trees. These efforts reveal that our galaxy closely resembles our closest large neighbor, the great spiral galaxy in Andromeda, with one notable difference: the Milky Way turns out to be a barred spiral galaxy. Conveniently located 2½ million light-years away, the Andromeda galaxy has provided a wealth of information about the basic structural patterns of spiral galaxies, as well as about different types of stars and their evolution. Because all of the Andromeda galaxy's stars have the same distance from us (plus or minus a few percent), astrophysicists know that the stars' brightnesses correlate directly with their luminosities, that is, with the amounts of energy they emit each second. This fact, denied to astrophysicists when they study objects in the Milky Way but applicable to every galaxy beyond our own, has allowed them to draw key conclusions about stellar evolution with greater ease than would be true for stars in the Milky Way. Two elliptical satellite galaxies that orbit the Andromeda galaxy, each containing a few percent of the number of stars in the main galaxy, have likewise furnished important information about the lives of stars, as well as the overall galactic structure of elliptical galaxies. On a clear night far from city lights, a keen-eyed observer who knows where to look can spot the fuzzy outline of the Andromeda galaxy—the most distant object visible to the unaided eye, shining with light that left on its journey while our ancestors roamed the gorges of Africa in search of roots and berries.

The Milky Way and the Andromeda galaxy each lie midway along their tines of Hubble's tuning-fork diagram, with spiral arms neither particularly tightly nor loosely wound. If galaxies were animals in a zoo, there would be one cage devoted to ellipticals but several animal houses for the glorious spirals. To study a Hubble Telescope image of one of these beasts, typically (for the closer ones) seen from 10 or 20 million light-years, is to enter a world of sight so rich in possibility, so deep in separation from life on Earth, so complex in structure, that the unprepared mind may reel, or may provide a defense by reminding its owner that none of this can thin the thighs or heal the fractured bone.

Irregulars, the orphans of the galactic class system, amount to about 10 percent of all galaxies, with the rest split between spirals and ellipticals, strongly favoring spirals. In contrast to ellipticals, irregular galaxies typically contain a higher proportion of gas and dust than spirals do, and offer the liveliest sites of ongoing star formation. The Milky Way has two large satellite galaxies, both irregular, confusingly named the Magellanic Clouds because the first Europeans to notice them, sailors on Magellan's circumnavigation of Earth in 1520, thought at first that they were seeing wisps of clouds in the sky. This honor fell to Magellan's expedition because the Magellanic Clouds lie so close to the south celestial pole (the point directly above Earth's South Pole) that they never rise above the horizon for observers in the most populated northern latitudes, including those in Europe and most of the United States. Each of the Magellanic Clouds contains many billion stars, though not the hundreds of billions that characterize the Milky Way and other large galaxies, and display immense star-forming regions, most notably the "Tarantula nebula" of the Large Magellanic Cloud. This galaxy also has the honor of having revealed the closest and brightest supernova to appear during the past three centuries, Supernova 1987A, which must have actually exploded about 160,000 BC for its light to have reached Earth in 1987.

Until the 1960s, astrophysicists were content to classify almost all galaxies as spiral, barred spiral, elliptical, or irregular. They had right on their side, since more than 99 percent of all galaxies fit one of these classes. (With one galactic class called "irregular," this result might seem to be a slam dunk.) But during that fine decade, an American astrophysicist named Halton Arp became the champion of galaxies that did not fit the simple classification scheme of the Hubble tuning-fork diagram plus irregulars. In the spirit of "Give me your tired, your poor, your huddled masses," Arp used the world's largest telescope, the 200-inch Hale Telescope at the Palomar Observatory near San Diego, California, to photograph 338 extremely disturbed-looking systems. Arp's *Atlas of Peculiar Galaxies*, published in 1966, became a veritable treasure chest of research opportunities on what can go bad in the universe.

Although "peculiar galaxies"—defined as galaxies with such strange shapes that even "irregular" fails to do them justice— form only a tiny minority of all galaxies, they carry important information about what can happen to galaxies gone wrong. It turns out, for example, that many embarrassingly peculiar galaxies in Arp's atlas are the merged remnants of two once-separate galaxies that have collided. This means that those "peculiar" galaxies are not different kinds of galaxies at all, any more than a wrecked Lexus is a new kind of car.

––––––––

To track how such a collision unfolds, you need a lot more than pencil and paper, because every star in both galactic systems has its own gravity, which simultaneously affects all the other stars in the two systems. What you need, in short, is a computer. Galaxy collisions are stately dramas, taking hundreds of millions of years from beginning to end. Using a computer simulation, you can start, and pause as you like, a collision of two galaxies, taking snapshots after 10 million years, 50 million years, 100 million years. At each time things look different. And when you step into Arp's atlas—bada-bing—here's an early stage of a collision, and there's a late stage. Here's a glancing blow, and there's a head-on collision.

Although the first computer simulations were done in the early 1960s (and although the Swedish astrophysicist Erik Holmberg made a clever attempt during the 1940s to recreate a galaxy collision on a tabletop by using light as an analogue to gravity), it wasn't until 1972 that Alar and Juri Toomre, brothers who both taught at MIT and the University of Colorado, generated the first compelling portrait of a "deliberately simple-minded" collision between two spiral galaxies.

The Toomres' model revealed that tidal forces—differences in gravity from place to place—actually rip the galaxies apart. As one galaxy nears the other, the gravitational force rapidly grows stronger at the leading edges of the collision, stretching and warping both galaxies as they pass by or through each other. That stretching and warping accounts for most of what's peculiar in Arp's atlas of peculiar galaxies.

How else can computer simulations help us to understand galaxies? Hubble's tuning fork distinguishes "normal" spiral galaxies from spirals like ours that show a dense bar of stars across their center. Simulations show that this bar could be a transitory feature, not the distinguishing mark of a different galactic species. Contemporary observers of barred spirals might simply be catching such galaxies during a phase that will disappear in 100 million years or so. But since we can't hang around long enough to watch the bar disappear in real life, we have to watch it come and go on a computer, where a billion years can unfold in a matter of minutes.

———

Arp's peculiar galaxies proved to be the tip of an iceberg, a strange world of not-exactly-galaxies whose outlines astrophysicists began to discern during the 1960s and came to understand a few decades later. Before we can appreciate this emergent galactic zoo, we must resume the story of cosmic evolution where we left it. We must examine the origin of all galaxies—normal, nearly normal, irregular, peculiar, and knock-your-socks-off exotic—to see how they were born, and how the luck of the draw has left us in our relatively calm location in space, adrift in the suburbs of a giant spiral galaxy, some 30,000 light-years from its center and approximately

20,000 light-years from its diffuse outer edge. Thanks to the
general order of things in a spiral galaxy, first imposed on the
gas clouds that later gave birth to stars, our Sun moves in a
nearly circular orbit around the center of the Milky Way, taking 240 million years (sometimes called a "cosmic year") for
each trip. Today, 20 orbits after its birth, the Sun should be
good for another 20 or so before calling it quits. Meanwhile,
let's have a look at where galaxies came from.

9

THE ORIGIN OF STRUCTURE

When we examine the history of matter in the universe, looking back through 14 billion years of time as best we can, we quickly encounter a single trend that cries out for explanation. Throughout the cosmos, matter has consistently organized itself into structures. From its nearly perfectly smooth distribution soon after the big bang, matter has clumped itself together on all size scales, to produce giant clusters and superclusters of galaxies, as well as the individual galaxies within those clusters, the stars that congregate by the billions in every galaxy, and quite possibly much smaller objects—planets, their satellites, asteroids, and comets—that orbit many if not most of those stars.

To understand the origin of the objects that now compose the visible universe, we must focus on the mechanisms that turned the universe's formerly diffuse matter into highly structured components. A complete description of how structures emerged in the cosmos requires that we meld two aspects of reality whose combination now eludes us. As seen in earlier chapters, we must perceive how quantum mechanics, which describes the behavior of molecules, atoms, and the particles that form them, fits with general relativity theory, which describes how extremely large amounts of matter and space affect one another.

Attempts to create a single theory that would unite our knowledge of the subatomically small and the astronomically large began with Albert Einstein. They have continued, with relatively little success, right up to the present time and will endure into an uncertain future, until they achieve "grand unification." Among all the unknowns that irk them, modern cosmologists feel most acutely the lack of a theory that triumphantly blends quantum mechanics with general relativity. Meanwhile, these seemingly immiscible branches of physics—the science of the small and the science of the large—care not a whit for our ignorance; instead, they coexist with remarkable success inside the same universe, whose various components mock our attempts to understand them from top to bottom as a coherent whole. A galaxy with 100 billion stars apparently pays no particular attention to the physics of the atoms and molecules that compose its star systems and gas clouds. Neither do the even larger agglomerations of matter we call galaxy clusters and superclusters, themselves containing hundreds, sometimes thousands, of galaxies. But these largest structures in the universe nonetheless owe their very existence to immeasurably small quantum fluctuations within the primeval cosmos. To understand how these structures arose, we must do the best we can in our current state of ignorance, passing from the minuscule domains governed by quantum mechanics, which hold the key to the origin of structure, to those so large that quantum mechanics plays no role, and matter obeys the laws laid down by general relativity.

To this end, we must seek to explain the structure-rich universe that we see today as arising from a nearly featureless cosmos soon after the big bang. Any attempt to explain the origin of structure must also account for the cosmos in its present state. Even this modest task has confounded astrophysicists and cosmologists with a series of false starts and errors, from which we have now (so we may fervently hope) removed ourselves to walk in the bright light of a correct description of the universe.

Throughout most of modern cosmology's history, astrophysicists have assumed that the distribution of matter in the universe can be described as both homogeneous and isotropic. In a homogeneous universe, every location looks similar to every other location, like the contents of a glass of homogenized milk. An isotropic universe is one that looks the same in every direction from any given point in space and time. These two descriptions may seem the same, but they are not. For example, the lines of longitude on Earth are not homogeneous, because they are farther apart in some regions and closer together in others; they are isotropic in just two places, the North and South Poles, where all lines of longitude converge. If you stand at either the "top" or "bottom" of the world, the longitude grid will look the same to you, no matter how far to the left or the right you turn your head. In a more physical example, imagine yourself atop a perfect, cone-shaped mountain, and that this mountain is the only thing in the world. Then every view of Earth's surface from that perch would be identical. The same would be true if you happened to live in the center of an archery target, or if you were a spider at the center of its perfectly spun web. In each of these cases, your view will be isotropic, but decidedly not homogeneous.

An example of a homogeneous but non-isotropic pattern appears in a wall of identical rectangular bricks, laid in a bricklayer's traditional overlapping manner. On the scale of several adjoining bricks and their mortar, the wall will be the same everywhere—bricks—but different lines of sight along the wall will intersect the mortar differently, destroying any claim to isotropy.

Intriguingly (for those who love a certain kind of intrigue), mathematical analysis tells us that space will turn out to be homogeneous only if it is everywhere isotropic. Another formal theorem of mathematics tells us that if space is isotropic in just three places, then space must be isotropic everywhere. Yet some of us shun mathematics as uninteresting and unproductive!

Although cosmologists were aesthetically motivated for assuming the homogeneity and isotropy of the distribution of matter in space, they have come to believe in this assumption enough to establish it as a fundamental cosmological principle. We might also call this the principle of mediocrity: Why should one part of the universe be any more interesting than another? On scales of size and distance familiar to us, we easily recognize this assertion to be false. We live on a solid planet with an average density of matter close to 5.5 grams per cubic centimeter (in Americanese, that's about 340 pounds per cubic foot). Our Sun, a typical star, has an average density of about 1.4 grams per cubic centimeter. The interplanetary spaces between the two objects, however, have a significantly smaller average density—smaller by a factor of about 1 billion trillion. Intergalactic space, which accounts for most of the volume of the universe, contains less than one atom in every 10 cubic meters, falling below the density of interplanetary space by another factor of 1 billion. Enough to make the mind feel good about the occasional accusation of being dense.

As astrophysicists expanded their horizons, they saw clearly that a galaxy such as our Milky Way consists of stars that float through nearly empty interstellar space. The galaxies likewise group into clusters that violate the assumption of homogeneity and isotropy. The hope remained, however, that as astrophysicists charted visible matter on the largest scales, they would find that galaxy clusters have a homogeneous and isotropic distribution. For homogeneity and isotropy to exist within a particular region of space, it must be large enough that no structures (or lack of structures) sit uniquely within it. If you take a melon-ball sample of such a region, the requirements of homogeneity and isotropy imply that the region's overall properties must be similar in every way to the average properties of any other scoop with the same size. What an embarrassment it would be if the left half of the universe looked different from its right half, no matter how we might define left and right.

How large a region must we examine to find a homogeneous and isotropic universe? Our planet Earth has a diameter of 0.04 light-seconds. Neptune's orbit spans 8 light-hours. The stars of the Milky Way galaxy delineate a broad, flat disk about 100,000 light-years across. And the Virgo supercluster of galaxies, to which the Milky Way belongs, extends some 60 million light-years. So the coveted volume that can give us homogeneity and isotropy must be larger than the Virgo supercluster. When astrophysicists made surveys of the galaxies' distribution in space, they discovered that even on these scales of size, as large as 100 million light-years, the cosmos reveals enormous, comparatively empty gaps, bounded by galaxies that have arranged themselves into intersecting sheets and filaments. Far from resembling a teeming, homogeneous anthill, the distribution of galaxies on this scale resembles a loofah sponge.

Finally, however, astrophysicists made still larger maps, and found their hoped-for homogeneity and isotropy. Turns out, a 300-million-light-year scoop of the universe does indeed resemble other scoops of the same size, fulfilling the long-sought aesthetic criterion for the cosmos. But, of course, on smaller scales, everything has clumped itself into distinctly nonhomogeneous and non-isotropic distributions of matter.

Three centuries ago, Isaac Newton considered the question of how matter acquired structure. His creative mind easily embraced the concept of an isotropic and homogeneous universe, but promptly raised an issue that would not occur to most of us: How can you make any structure at all in the universe without having all the matter of the universe joining it to create one gigantic mass? Newton argued that since we observe no such mass the universe must be infinite. In 1692, writing to Richard Bentley, the master of Trinity College at Cambridge University, Newton proposed that

> if all the matter in the universe were evenly scattered through-out all the heavens, and every particle had an innate gravity toward all the rest, and the whole space throughout which this matter was scattered was but finite, the matter on the outside of the space would, by its gravity, tend toward all the matter on the inside, and by consequence, fall down into the middle of the whole space and there compose one great spherical mass. But if the matter was evenly disposed throughout an infinite space, it could never convene into one mass; but some of it would convene

into one mass and some into another, so as to make an infinite number of great masses, scattered at great distances from one to another throughout all that infinite space.

Newton presumed that his infinite universe must be static, neither expanding nor contracting. Within this universe, objects were "convened" by gravitational forces—the attraction that every object with mass exerts on all other objects. His conclusion about gravity's central role in creating structure remains valid today, even though cosmologists face a task more daunting than Newton's. Far from enjoying the benefits of a static universe, we must allow for the fact that the universe has been expanding ever since the big bang, naturally opposing any tendency for matter to clump together by gravity. The problem of overcoming the cosmic expansion's anticonvening tendency becomes more serious when we consider that the cosmos expanded most rapidly soon after the big bang, the era when structures first began to form. At first glance, we could no more rely on gravity to form massive objects out of diffuse gas than we could use a shovel to move fleas across a barnyard. Yet somehow gravity has done the trick.

During the early days of the universe, the cosmos expanded so rapidly that if the universe had been strictly homogeneous and isotropic on all size scales, gravity would have had no chance of victory. Today there would be no galaxies, stars, planets, or people, only a scattered distribution of atoms everywhere in space—a dull and boring cosmos, devoid of admirers and objects of admiration. We live within a fun and exciting universe only because *in*homogeneities and *ani*sotropies appeared during those earliest cosmic moments, which served as a kind of cosmic soup-starter for all concentrations of matter and energy that would later emerge. Without this head start, the rapidly expanding universe would have prevented gravity from ever gathering matter to build the familiar structures we take for granted in the universe today.

What made these deviations, the inhomogeneities and anisotropies that provide the utterly essential seeds for all the structure in the cosmos? The answer arrives from the realm of quantum mechanics, undreamt of by Isaac Newton but essential if we hope to understand where we came from. Quantum mechanics tells us that on the smallest scales of size, no distribution of matter can remain homogeneous and isotropic. Instead, random fluctuations in the distribution of matter will appear, disappear, and reappear in different amounts, as matter becomes a quivering mass of vanishing and reborn particles. At any particular time, some regions of space will have slightly more particles, and therefore a slightly greater density, than other regions. From this counterintuitive, Swiss-cheese fantasy, we derive everything that exists. The slightly denser regions had the chance to attract slightly more particles by gravity, and through the passage of time the cosmos grew these denser regions into the structures that we see today.

In tracing the growth of structure from times soon after the big bang, we can gain some insight from two key epochs we have already met, the extremely early "era of inflation," when the universe expanded at an astounding rate, and the later "time of decoupling," still quite early, about 380,000 years after the big bang, when the cosmic background radiation ceased to interact with matter.

The inflationary era lasted from about 10^{-37} second to 10^{-33} second after the big bang. During that relatively brief stretch of time, the fabric of space and time expanded faster than light, growing in a billionth of a trillionth of a trillionth of a second from 100 billion billion times smaller than the size of a proton to about four inches. Yes, the observable universe once fit within a grapefruit. But what caused this cosmic inflation? Cosmologists have named the culprit: a "phase transition" that left behind a specific and observable signature in the cosmic background radiation.

Phase transitions are hardly unique to cosmology; they often occur in the privacy of your home. We freeze water to make ice cubes, and boil water to produce steam. Sugary water grows sugar crystals on a string dangling within the liquid. And wet, gooey batter turns into cake when baked. There's a pattern here. In every case, things look very different on the two sides of a phase transition. The inflationary model of the universe asserts that when the universe was young, the prevailing energy field went through a phase transition, one of several that would have occurred during these early times. This particular episode not only catapulted the early, rapid expansion but also imbued the cosmos with a specific fluctuating pattern of high- and low-density regions. These fluctuations then froze into the expanding fabric of space, creating a kind of blueprint for where galaxies would ultimately form. Thus, in the spirit of Pooh-Bah, the character in Gilbert and Sullivan's *Mikado* who proudly traced his ancestry back to a "primordial atomic globule," we can assign our origins, and the beginnings of all structure, to the fluctuations on a subnuclear scale that arose during the inflationary era.

What facts can we cite to support this bold assertion? Since astrophysicists have no way to see back to the universe's first 0.000000000000000000000000000000000001 of a second, they do the next best thing, and use scientific logic to connect this early epoch to times they can observe. If the inflationary theory is correct, the initial fluctuations produced during that era, the inevitable result of quantum mechanics—which tells us that small variations from place to place will always arise within an otherwise homogeneous and isotropic fluid—would have had the opportunity to become regions of high and low concentrations of matter and energy. We can hope to find evidence for these variations from place to place in the cosmic background radiation, which serves as a proscenium that separates the current epoch from, and also connects it with, the first moments of the neonate universe.

As we have already noted with admiration, the cosmic background radiation consists of the photons generated during the first minutes after the big bang. Early in the universe's history, these photons interacted with matter, slamming so energetically into any atoms that happened to form that no atoms could exist for long. But the ceaseless expansion of the universe effectively robbed the photons of energy, so that eventually, at the time of decoupling, none of the photons had energies sufficient to prevent electrons from orbiting around protons and helium nuclei. Since that time, 380,000 years after the big bang, atoms have persisted—unless some local disturbance, such as the radiation from a nearby star, disrupts them— while the photons, each with an ever-diminishing amount of energy, continue to roam the universe, collectively forming the cosmic background radiation, or CBR.

The CBR thus carries the imprint of cosmic history, a snapshot of what the universe was like at the time of decoupling. As we described in previous chapters, astrophysicists have learned how to examine this snapshot with ever-increasing accuracy. First, the fact that the CBR exists demonstrates that their basic understanding of the history of the universe is correct. And then, after years of improving their abilities to measure the cosmic background radiation, their sophisticated balloon-borne and satellite instruments gave them a map of the CBR's tiny deviations from homogeneity. This map provides the record of the once-minuscule fluctuations, whose size increased as the universe expanded during the few hundred thousand years after the era of inflation, and which then grew, during the next billion years or so, into the large-scale distribution of matter in the cosmos.

Remarkable though it may seem, the CBR provides us with the means for mapping the imprint of the long-vanished early universe, and for locating—14 billion light-years away in all directions—the regions of slightly greater density that would become galaxy clusters and superclusters. Regions with greater-than-average density left behind slightly more photons than regions with lower densities. As the cosmos became transparent, thanks to the loss of energy that left the photons unable to interact with the newly formed atoms, each photon embarked on a journey that would carry it far from its point of origin. Photons from our vicinity have traveled almost 14 billion light-years in all directions, providing part of the CBR that far-distant civilizations at the end of the visible universe may even now be examining, and "their" photons, having reached our instruments, tell us about what things were like long ago and far away, in the times when structures had barely begun to form.

Through more than a quarter of a century following the first detection of the cosmic background radiation in 1965, astrophysicists searched for anisotropies in the CBR. From a theoretical viewpoint, they desperately needed to find them, because without the existence of CBR anisotropies at the level of a few parts in a hundred thousand, their basic model of how structure appeared would lose any claim to validity. Without the seeds of matter they reveal, we would have no explanation for why we exist. As happy fate would have it, the anisotropies appeared precisely on schedule. Just as soon as cosmologists created instruments capable of detecting anisotropies at the appropriate level, they found them, first with the COBE satellite in 1992, and later with far more precise instruments mounted on balloons and on the WMAP satellite described in Chapter 3. The teeny fluctuations from place to place in the amounts of microwave photons that form the CBR, delineated with impressive precision by WMAP and still more precisely by Planck, embody the record of cosmic fluctuations at a time 380,000 years after the big bang. Typical fluctuations sit only a few hundred thousandths of a degree above or below the average temperature of the cosmic background radiation, so detecting them is like finding faint spots of oil on a mile-wide pond that make the water plus oil a shade less dense than average. Small though these anisotropies were, they sufficed to get things started.

In modern maps of the cosmic background radiation, the larger hot spots tell us where gravity would overcome the expanding universe's dissipative tendencies and gather together enough matter to manufacture superclusters. These regions today have grown to contain about 1,000 galaxies, each with 100 billion stars. If we add the dark matter in such a supercluster, its total mass reaches the equivalent of 10^{16} Suns. Conversely, the larger cool spots, with no head start against the expanding universe, evolved into regions nearly

In July 2022, NASA released its first image from the James Webb Space Telescope. Unlike the much smaller Hubble Space Telescope, the JWST can detect infrared light, and avoids the Earth's glare by orbiting the sun at four times the moon's distance. In this "deep field" image, a tiny region of the sky (equal in angular size to a grain of sand held at arm's length) is dominated by a galaxy cluster whose light left at the same time that the Earth formed, 4.6 billion years ago. Almost every dot in the image is a galaxy containing billions of stars. Gravitational forces from some of them have bent the light from still more distant galaxies, as far as 13 billion light-years away, to produce curved arcs.

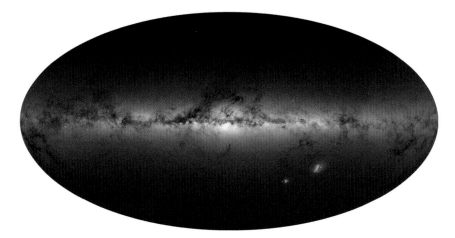

The Gaia satellite, created and operated by the twenty-two countries that form the European Space Agency, took almost two years to produce the most complete and most accurate star catalog, with approximately 1.7 billion stellar positions and motions. This all-sky map shows the dominance of the sky by the "milky way," the regions close to the central plane of our galaxy, where billions of stars shine amid interstellar dust that absorbs some of their light.

The Planck satellite, created by the European Space Agency, spent the years 2009 to 2013 mapping the photons released during the first 400,000 years after the big bang, which have been traveling through space ever since then, steadily weakened by the universe's expansion. Slight differences from place to place in the strength of this cosmic background radiation, coded here as red for slightly stronger and blue for slightly weaker, reveal important facts about the universe. They confirm that most of the matter in the universe is "dark matter," and most of the energy is "dark energy." In addition, these observations suggest a rate of cosmic expansion slightly different from that derived from observations of supernovae, creating an unreconciled conflict, known among cosmologists as "cosmic tension."

In this image of the Coma cluster of galaxies, 325 million light-years from the Milky Way, almost every smudge of light is in fact a galaxy composed of billions of stars. The thousands of galaxies execute a complex ballet around the center of mass of the cluster, which spans several million light-years.

The closest large cluster of galaxies, the Virgo cluster, lies a mere 60 million light-years from the Milky Way. Among the cluster's more than a thousand members are huge elliptical galaxies, visible at the upper left and lower center, and the giant spiral galaxy at the upper right. Keen eyes can spot many dozen galaxies in this image. The Milky Way and its small Local Group of galaxies belong to the outer fringes of this cluster, known in its entirety as the Virgo *super*cluster.

The giant cluster of galaxies A2218 lies about 3 billion light-years from the Milky Way. Behind the galaxies in this cluster lie still-more-distant galaxies, whose light is bent and distorted by the gravitational forces from the galaxies and dark matter in the cluster. The bending produces the large, thin arcs of light in this image obtained by the Hubble Space Telescope.

Galaxies in the giant cluster A1689, 2 billion light-years away, likewise bend light from more-distant galaxies. Measurement of the details of the arcs of light produced by the bending allowed astrophysicists to determine that most of the cluster's mass resides not in the galaxies themselves but rather in dark matter surrounding the visible galaxies.

The quasars shown here are 3C 273, fully 2.4 billion light-years from the Milky Way but nevertheless one of the closest quasars, and PKS 1127-145, four times farther away at 10 billion light-years. The Hubble Space Telescope secured the image of 3C 273, while the Chandra telescope obtained the X-ray image of PKS 1127-145. Both images show the quasar ejecting material in a long, thin jet; in the case of PKS 1127-145, the jet extends for a million light-years. The sharp spikes around the image of 3C 273 are optical effects that arise in the telescope itself.

One of the largest galaxies in the Virgo cluster, catalogued as NGC 4535, has comparatively thin spiral arms, composed of stars plus gas and dust. The galaxy lies on the "near side" of the cluster, about 54 million light-years from the Milky Way.

The gravitational forces between this peculiar pair of galaxies, located about 325 million light-years from the Milky Way, called Arp 295 from their numbering in Halton Arp's *Atlas of Peculiar Galaxies*, have drawn long streams of matter across the quarter of a million light-years of space between them.

The giant Pinwheel galaxy, best known as M101, lies 21 million light-years away in the direction of Ursa Major. It has almost twice the Milky Way's diameter and contains at least a trillion stars. This image is a mosaic of 51 separate exposures made by the Hubble Space Telescope. Observations of large star-forming regions and giant clouds of molecular hydrogen (hydrogen atoms joined in pairs) allow astrophysicists to trace the galaxy's spiral arms in detail.

The spiral galaxy NGC 3370, about 100 million light-years from us and 100,000 light-years across, closely resembles our own Milky Way in its size, shape, and mass. This image from the Hubble Space Telescope shows the complex spiral distributions traced by young, highly luminous, hot stars as well as the star-forming complexes still creating new stars.

In March 1994, the Hubble Space Telescope obtained this image of the supernova that appeared in the outer regions of NGC 4526, one of the galaxies in the Virgo cluster, about 60 million light-years from the Milky Way. Seen nearly edge-on, the galaxy shows the absorption of light by the bands of dust that lie close to the galaxy's central plane. This Type Ia supernova belongs to the class of exploding stars that astrophysicists used to discover the universe's accelerating rate of expansion.

This image of the spiral galaxy NGC 4256, about 23 million light-years from the Milky Way, combines observations in X-rays (shown in blue), radio waves (purple), visible light (yellow and blue), and infrared (in red). Unlike most spirals, this galaxy has additional, anomalous spiral arms that extend above and below its central plane. Astrophysicists deduce that a supermassive black hole at the galaxy's center strongly affects the motions of the gases that surround it.

Dwarf Galaxy NGC 1569
Hubble Space Telescope • Wide Field Planetary Camera 2

ESA, NASA & P. Anders (Göttingen University, Germany)　　　　　　　　　　　　　　　STScI-PRC04-06

The dwarf galaxy NGC 1569, about 11 million light-years from the Milky Way, displays the results of intense star-birth activity that began about 25 million years ago and persisted through the next 20 million years. In this stretch of time, the most massive stars have produced supernovae that have blown "bubbles" through the galaxy.

The Milky Way has two large satellite galaxies, the Large and Small Magellanic Clouds, about 160,000 light-years away. Visible primarily from the southern hemisphere, these are named for Ferdinand Magellan, who described them on his round-the-world voyage five centuries ago. This image of the Large Magellanic Cloud shows a large bar of stars at the left, with many additional stars and a huge star-forming region to the right.

The Large Magellanic Cloud's great star-forming region, the "Tarantula nebula," spans almost a thousand light-years. "Super star clusters" that contain high concentrations of young, hot, massive stars create the energy that makes the gases glow.

The constellation Centaurus, visible in the southern regions of the sky, contains the giant glob- ular star cluster known as Omega Centauri, located about 17,000 light-years from the solar sys- tem. This image, which reveals about 300,000 of its 10 million stars, was secured at the Paranal Observatory of the European Southern Observatory organization, which has 16 member states across Europe.

M31

HST
PHAT

DSS and R. Gendler

The closest large galaxy to the Milky Way, about 2.2 million light-years away, lies in the direction of the constellation Andromeda—so close that its outermost regions span ten times the Moon's diameter in the sky. This image, largely created from observations by the Hubble Space Telescope, has been assembled by using 7,398 exposures taken during 411 separate pointings of the telescope toward the galaxy. Young blue stars appear predominantly beyond the central regions, where older yellowish stars dominate.

From observations made using two different bands of light, the Hubble Space Telescope obtained these two images of the central portion of the Eagle nebula that scientists named the "Pillars of Creation." The visible-light image on the left shows opaque clouds of gas and dust, within which new stars have formed and are still forming. As shown in the right-hand image, the shorter-wavelength type of infrared penetrates much of this material to reveal stars whose visible light has been absorbed by interstellar dust.

devoid of massive structures. Astrophysicists just call these regions "voids," a term that gains meaning from their being surrounded by something that is not a void. So the giant sheets and filaments of galaxies that we can trace on the sky not only form clusters at their intersections but also trace walls and other geometric forms that give shape to the empty regions of the cosmos.

Of course, the galaxies themselves did not simply appear, fully formed, from concentrations of matter a tiny bit denser than average. From 380,000 years after the big bang until about 200 million years later, matter continued to gather itself together, but nothing shone in the universe, whose first stars were yet to be born. During this cosmic dark age, ordinary matter in the universe incorporated the elements that it had made during its first few minutes—hydrogen and helium, with traces of lithium. With no elements heavier than these— no carbon, nitrogen, oxygen, sodium, calcium, or heavier elements—the cosmos contained none of the now-common atoms or molecules that can absorb light as a star begins to shine. Today, in the presence of these atoms and molecules, the light from a newly formed star will exert pressure upon them that pushes away massive quantities of gas that would otherwise fall into the star. This expulsion limits the maximum mass of newborn stars to less than a hundred times the Sun's mass. But when the first stars formed, in the absence of atoms and molecules that would absorb starlight, infalling gas consisted almost entirely of hydrogen and helium, providing only token resistance to stars' output. This allowed stars to form with much larger masses, up to many hundred, perhaps even a few thousand, times the mass of the Sun.

High-mass stars live life in the fast lane, and the most massive live the most rapidly of all. They convert their matter into energy at astonishing rates, as they manufacture heavy elements and die explosive youthful deaths. Their life expectancies amount to no more than a few million years, less than a thousandth of the Sun's. We expect to find none of the most massive stars from that era alive today, because the early ones burnt themselves out long ago, and today, with heavier elements common throughout the universe, the highest-mass stars of old cannot form at all. Indeed, none of the high-mass giants has ever been observed. Nevertheless, we assign them responsibility for having first introduced into the universe almost all of the familiar elements we now take for granted, including carbon, oxygen, nitrogen, silicon, and iron. Call it enrichment. Call it pollution. But the seeds of life began with the long-vanished first generation of high-mass stars.

―――――――

During the first few billion years after the time of decoupling, gravitationally induced collapse proceeded with abandon, as gravity drew matter together on nearly all scales. One of the natural results of gravity at work was the formation of supermassive black holes, each with a mass millions or billions of times the mass of the Sun. Black holes with that amount of mass are about the size of Neptune's orbit and wreak havoc on their nascent environment. Gas clouds drawn toward these black holes want to gain speed, but they can't, because there's too much stuff in the way. Instead, they slam into and rub against whatever came in just before them, descending toward their master in a swirling maelstrom. Just before these clouds disappear forever, collisions within their superheated matter radiate tremendous quantities of energy, billions of times the

Sun's luminosity, all within the volume of a solar system. Monstrous jets of matter and radiation spew forth, extending hundreds of thousands of light-years above and below the swirling gas, as the energy punches through and escapes the funnel in all ways it can. As one cloud falls, and another orbits in waiting, the luminosity of the system fluctuates, getting brighter and dimmer over a matter of hours, days, or weeks. If the jets happen to be aimed straight at you, the system will look even more luminous, and more variable in its output, than those cases in which the jets point to the side. Viewed from any appreciable distance, all of these black-hole-plus-infalling-matter combinations will appear amazingly small and luminous in comparison to the galaxies we see today. What the universe has created—the objects whose birth we have just witnessed in words—are quasars.

Quasars were discovered during the early 1960s, as astrophysicists began to perfect telescopes equipped with detectors sensitive to invisible domains of radiation, such as radio waves and X-rays. Their galaxy portraits could therefore include information about the galaxies' radiation output and appearance in those other bands of the electromagnetic spectrum. Combine this with further improvements in photographic emulsions, and a new zoo of galaxy species emerged from the depths of space. Most remarkable among them were objects that, in photographs, look like simple stars but—quite unlike stars—produce extraordinary amounts of energy in the form of radio waves. The working description for those objects was "quasistellar radio source"—a term quickly shortened to "quasar." Even more remarkable than the radio emission from these objects were their distances: as a class, they turned out to be the most distant objects known in the universe. For quasars to be that small and still visible at immense distances meant that they had to be an entirely new kind of object. How small? No bigger than a solar system. How luminous? Even the dim ones outshine your average galaxy throughout the cosmos.

By the early 1970s, astrophysicists had converged on supermassive black holes as the quasar engine, gravitationally devouring everything in its grasp. The black hole model can account for how small and bright quasars are, but says nothing of the black hole's source of food. Not until the 1980s would astrophysicists begin to understand the quasar's environment, because the tremendous luminosity of a quasar's central regions prevents any sight of its much fainter surroundings. Eventually, however, with new techniques to mask the light from the center, astrophysicists could detect fuzz surrounding some of the dimmer quasars. As detection tactics and technologies improved further, every quasar revealed fuzz; some even revealed a spiral structure. Quasars, it turned out, are not a new kind of object but rather a new kind of galactic nucleus.

———

In April 1990, the National Aeronautics and Space Administration (NASA) launched the most expensive astronomical instrument ever built: the Hubble Space Telescope. The size of a Greyhound bus, directed by commands sent from Earth, the Hubble Telescope could profit from orbiting outside our ever-blurring atmosphere. Once astronauts had installed lenses to correct for the incorrect shape that had been given to its primary mirror, the telescope could peer into previously uncharted regions of ordinary galaxies, including their centers. Upon gazing into those centers, it found stars moving inexplicably rapidly, given the gravity inferred from the visible light of other stars in the vicinity. Hmmm, strong gravity, small area . . . must be a black hole. Galaxy after galaxy— dozens of them—had suspiciously speedy stars in their cores. Indeed, whenever the Hubble Space Telescope had a clear view of a galaxy's center, they were there.

It now seems likely that every giant galaxy harbors a supermassive black hole, which could have served as a gravitational seed around which the other matter collected, or may have been manufactured later by matter streaming down from outer regions of the galaxy. But not all galaxies were quasars in their youth.

––––––

The growing roster of ordinary galaxies known to have a black hole at their center began to raise eyebrows among investigators: A supermassive black hole that was not a quasar? A quasar that's surrounded by a galaxy? One can't help but think of a new picture of how things work. In this picture, some galaxies begin their lives as quasars. To be a quasar, which is really just the blazing visible core of an otherwise run-of-the-mill galaxy, the system has to have not only a massive, hungry black hole but also an ample supply of infalling gas. Once the supermassive black hole has gulped down all the available food, leaving uneaten stars and gas in distant, safe orbits, the quasar simply shuts off. You've then got a docile galaxy with a dormant black hole snoozing at its center. Astrophysicists have found other new types of objects, classified as intermediate between quasars and normal galaxies, whose properties also depend on the bad behavior of supermassive black holes. Sometimes the streams of material falling into a galaxy's central black hole flow slowly and steadily. At other times episodically. Such systems populate the menagerie of galaxies whose nuclei are active but not ferocious. Over the years, names for the various types accumulated: LINERs (low-ionization nuclear emission-line regions), Seyfert galaxies, N galaxies, blazars. All of these objects are generically called AGNs, the astrophysicist's abbreviation for galaxies with "active" nuclei. Unlike quasars, which appear only at immense distances, AGNs appear both at large distances and relatively nearby. This suggests that AGNs fill in the range of galaxies that misbehave. Quasars long ago consumed all their food, so we see them only when we look far back in time by observing far out in space. AGNs, in contrast, had more modest appetites, so some of them still have food to eat even after billions of years.

Classifying AGNs on the basis of their visual appearance alone would provide an incomplete story, so astrophysicists classified AGNs by their spectra and over the full range of their electromagnetic emissions. During the mid- to late 1990s, investigators improved their black hole model, and found that they could characterize nearly all the beasts in the AGN zoo by measuring only a few parameters: the mass of the object's black hole, the rate at which it's being fed, and our angle of view on the accretion disk and its jets. If, for example, we happen to look "right down the barrel," along exactly the same direction as that of a jet emerging from the vicinity of a supermassive black hole, we see a much brighter object than if we happen to have a side view from a much different angle. Variations in these three parameters can account for nearly all the impressive diversity that astrophysicists observe, giving them a welcome de-speciation of galaxy types and a deeper understanding of the formation and evolution of galaxies. The fact that so much can be accounted for—differences in shape, size, luminosity, and color—by so few variables represents an unheralded triumph of late twentieth-century astrophysics. Because it took a lot of investigators and a lot of years and a lot of telescope time, it's not the sort of thing that gets announced on the evening news—but it's a triumph nonetheless.

———

Let us not conclude, however, that supermassive black holes can explain everything. Even though they each have millions or billions of times the Sun's mass, they contribute almost nothing in comparison with the masses of the galaxies in which they are embedded—typically far less than 1 percent of a large galaxy's total mass. When we seek to account for the existence of dark matter, or of other unseen sources of gravity in the universe, these black holes are insignificant and may be ignored. But when we calculate how much energy they wield—that is, when we compute the energy that they released as part of their formation—we find that black holes dominate the energetics of galaxy formation. All the energies of all the orbits of all the stars and gas clouds that ultimately compose a galaxy pale when compared with what made the black hole. Without supermassive black holes lurking below, galaxies as we know them might have never formed. The once-luminous-but-now-invisible black hole that lies at the center of each giant galaxy provides a hidden link, the physical explanation for the agglomeration of matter into a complex system of billions of stars in orbit around a common center.

The broader explanation for the formation of galaxies invokes not only the gravity produced by supermassive black holes but also gravity in more conventional astronomical settings. What made the billions of stars in a galaxy? Gravity did this too, producing up to hundreds of thousands of stars in a single cloud. Most of a galaxy's stars were born within relatively loose "associations." The more compact regions of star birth still remain as identifiable "star clusters," within which member stars orbit the cluster's center, tracing their paths through space in a cosmic ballet choreographed by the forces of gravity from all the other stars within the cluster, even as the clusters themselves move on enormous trajectories around the galactic center, safe from the destructive power of the central black hole.

Within a cluster, stars move at a broad range of speeds, some so rapidly that they risk escaping from the system altogether. This indeed occasionally occurs, as fast-moving stars evaporate from the grip of a cluster's gravity to roam freely through the galaxy. These free-ranging stars, along with the "globular star clusters" that contain hundreds of thousands of stars each, add to the stars that form the spherical haloes of galaxies. Initially luminous, but today devoid of their brightest, short-lived stars, galaxy haloes are the oldest visible objects in the universe, with birth certificates traceable to the time of formation of galaxies themselves.

The last to collapse, and thus the last to turn into stars, are the clouds of gas and accompanying dust that find themselves pulled and pinned into the galactic plane, where the gas has an increased chance of forming stars. In elliptical galaxies, no such plane exists, and all of their gas has already turned into stars. Spiral galaxies, however, have highly flattened distributions of matter, characterized by a central plane within which the youngest, brightest stars form in spiral patterns, testimony to great vibrating waves of alternating dense and rarefied gas that orbit the galactic center. Like hot marsh-mallows that stick together upon contact, all of the gas in a spiral galaxy that did not swiftly participate in making star clusters has fallen toward the galactic plane, stuck to itself, and created a disk of matter that slowly manufactures stars. Through billions of years, and for billions of years to come, stars have formed and will continue to form in spiral galaxies, with each generation more enriched in heavy elements than the last. These heavy elements (by which astrophysicists mean all elements heavier than helium) have been cast forth into interstellar space by outflows from aging stars or as the explosive remains of high-mass stars, a species of supernova. Their existence renders the galaxy—and thus the universe—ever friendlier to the chemistry of life as we know it.

We have outlined the birth of a classical spiral or barred spiral galaxy, in an evolutionary sequence that has played out tens of billions of times, yielding galaxies in an abundance of different arrangements: In clusters of galaxies. In long strings and filaments of galaxies. And in sheets of galaxies.

Because we look back in time as we look outward into space, we possess the ability to examine galaxies not only as they are now but also as they appeared billions of years ago, simply by looking outward. The difficulty with turning this concept into observational reality resides in the fact that galaxies billions of light-years away appear to us as extremely small and dim objects, so even our best telescopes can barely resolve their outlines. Nevertheless, astrophysicists have made great progress in this effort during the past few years. A key breakthrough came in 1995, when Robert Williams, then the director of the

Space Telescope Science Institute at Johns Hopkins University, arranged for the Hubble Telescope to point toward a single direction in space, near the Big Dipper, for 10 days' worth of observation. Williams deserves the credit because the telescope's Time Allocation Committee, which selects the observing proposals most worthy of actual telescope time, judged it unworthy of support. After all, the region to be studied was deliberately chosen for having nothing interesting to look at, and thus to represent a dull and boring patch of sky. As a result, no ongoing projects could benefit directly from such a large commitment of the telescope's highly oversubscribed observing time. Fortunately, Williams, as the director of the Space Telescope Science Institute, had the right to assign a few percent of the total—his "director's discretionary time"— and invested his clout on what became known as the Hubble Deep Field, one of the most famous astronomical photographs ever taken.

The 10-day exposure, coincidentally made during the government shutdown of 1995, produced, by far, the most researched image in the history of astronomy. Studded with galaxies and galaxy-like objects, the deep field offers us a cosmic palimpsest, in which objects at different distances from the Milky Way have written their momentary signatures of light at different times. We see objects in the deep field as they were, say, 1.3 billion, 3.6 billion, 5.7 billion, or 8.2 billion years ago, with each object's epoch of visibility determined by its distance from us. Hundreds of astrophysicists have seized upon the wealth of data contained in this single image to derive new information about how galaxies have evolved with time, and about how galaxies looked soon after they formed. In 1998, the telescope secured a companion image, the Hubble Deep Field South, by devoting 10 days of observation to another patch of sky in the direction opposite to that of the first deep field, in the celestial southern hemisphere. Comparison of the two images allowed astrophysicists to assure themselves that the results from the first deep field did not represent an anomaly (for example, if the two images had been identical in every detail, or statistically unlike each other in every way, one might have concluded that the devil was at work), and to refine their conclusions about how different types of galaxies form. After a successful servicing mission, in which the Hubble Telescope was outfitted with even better (more sensitive) detectors, the Space Telescope Science Institute took the obvious next step: in 2004, the telescope secured the Hubble Ultra Deep Field, laying bare the farthest reaches of the cosmos that we can hope to see in visible light.

Unfortunately, the earliest stages of galaxy formation, which would be revealed to us from observations of objects at the greatest distances from us, have confounded even the Hubble Telescope's best efforts, primarily because the cosmic expansion has shifted most of the radiation from these objects into the infrared region of the spectrum, which the Hubble's instruments cannot detect. For these most distant galaxies, astrophysicists long awaited the Hubble's successor, the James Webb Space Telescope (JWST), named after the head of NASA during the Apollo era and launched on Christmas Day 2021.

The JWST has a mirror 2½ times wider than Hubble's, made not from a single great piece of glass but from a set of 18 hexagonal mirrors, designed to unfurl themselves in space like an intricate mechanical flower, then snap together meticulously to become a near-perfect reflective surface much larger than any that can fit inside one of our launch vehicles. The new space telescope also carries a suite of instruments far superior to those of the Hubble Telescope, which were originally designed during the 1960s, built during the 1970s, launched in 1991, and—even though significantly upgraded during the 1990s—continued to lack such fundamental abilities as the capacity to detect infrared radiation. Some of this ability existed in the modest-sized Spitzer Space Telescope, launched in 2003 and retired in 2020, which orbited the Sun much farther from Earth than the Hubble does, thereby avoiding interference from the copious amounts of infrared radiation produced by our planet.

To achieve the same avoidance of the glare from Earth, the JWST has likewise reached an orbit much farther from Earth than the Hubble Telescope's. At "L2," the point in space in a direction exactly opposite to the Sun, at almost four times the Moon's distance from Earth, the JWST can easily maintain a constant position with respect to Earth as both the telescope and our planet orbit the Sun. The L2 point's million-mile distance renders it inaccessible with our current servicing mission capabilities, so NASA had to get this one right the first time. Following months of testing, the JWST began full operations in mid-2022, ready to employ its specialized capabilities and mammoth size to provide humanity with new insights into the cosmos.

Those insights will primarily spring from the JWST's ability to observe the infrared universe, a capability sorely and sadly lacking in the Hubble's suite of instruments. Objects observed at distances of many billion light-years have had the wavelengths of their light increased by factors of 5, 10, or even 20 or more. Radiation emitted as visible light, and even much of their ultraviolet output, has been shifted well into the domain of the infrared, requiring specialized detectors for detailed study. The JWST's infrared capabilities will allow it to observe the epoch of galaxy formation, which began less than a billion years after the big bang.

———

In addition to these improvements in now-basic methods for astronomical observation, the past few years have brought forth a new means of cosmic exploration. In 2015 an entirely new observational window opened with the first direct detection of "gravitational radiation."

Among scientists, "radiation" carries several meanings. Usually the word describes electromagnetic radiation, streams of massless photons, of which different types carry different amounts of energy per photon. The confusing term "nuclear radiation" includes both photons and particles with mass that are involved in nuclear reactions. Entirely different from both of those, gravitational radiation, also called gravitational waves, is ripples of space itself that make anything that they encounter wiggle in response to their passage. Photons travel *through* space, but gravitational waves *are* space—more precisely, the rhythmic bending of space that spreads outward at the speed of light. If gravitational waves were not so feeble— a reminder that gravity ranks as by far the weakest of nature's basic forces—we would feel ourselves, along with our planet, being constantly jostled by intense gravitational radiation, produced by the most violent events in the universe that make space quiver.

Their weakness made gravitational waves wait nearly a century to pass from prediction to detection. In 1916, applying his general theory of relativity's insistence that gravitational forces bend space, Einstein concluded that a source of strong gravitational force, moving extremely rapidly, could create detectable waves in space. These ripples would spread outward, losing strength just as light waves do. In theory, an entirely new sort of instrument could record them independently of any electromagnetic radiation that might have emerged from the event.

Einstein's analysis noted the extreme weakness of the gravitational waves implicit in his theory. He doubted that direct detection could occur, a judgment that remained valid for generations. During the final decades of the twentieth century, astrophysicists discovered systems in which two "neutron stars," the collapsed cores of stars that have exploded as supernovae, produce regular pulses of radio waves as they orbit their common center of mass. The neutron stars move in such close proximity, and at such high speeds, that gravitational waves carry away energy in amounts sufficient to shrink the sizes of their orbits. The metronomic regularity of the radio pulses allowed measurement of the changing sizes of the neutron stars' orbits from tiny changes in the radio pulses' arrival times. These measurements verified Einstein's prediction of gravitational radiation by deduction rather than direct detection, and led to Russell Hulse and Joseph Taylor sharing the 1993 Nobel Prize in physics.

Two decades passed before a system capable of directly recording Einstein's ripples in space began operation. Success required the construction of identical detectors in Louisiana and Washington State that together form LIGO, the Laser Interferometer Gravitational-Wave Observatory. In each detector, two 4-kilometer-long tunnels form a giant L. At the ends of these arms, mirrors reflect laser beams back to other mirrors where the arms join, so that the beams travel back and forth through a near-perfect vacuum hundreds of times before they meet for comparison at the center point of the L. This configuration allows LIGO's scientists to detect differences as small as 10 billion-billionths of a centimeter between the total lengths of the paths that the laser beams have traveled. By deploying two detectors separated by thousands of miles, they can ensure that any such tiny differences, recorded almost simultaneously, must arise from actual ripples of space rather than from nearby disturbances. In 2018, after an upgrade that improved the detectors' sensitivity, LIGO recorded its first gravitational waves. A third detector, named VIRGO, located near Pisa, Italy, and a fourth, KAGRA, in the center of Japan, have joined what has become a worldwide effort, a welcome result that allows scientists to use tiny time differences in the passage of the ripples, which arise from the finite speed of light, to determine the direction from which the gravitational waves arrive. The 2017 physics Nobel Prize recognized Rainer Weiss and Kip Thorne, experts in the theory of gravitational radiation, and Barry Barish, the key figure in LIGO's construction.

The chief sources of gravitational waves (at least on human time scales) involve the collision and merger of massive condensed objects. Years of theoretical research have given the now hundreds-strong community of gravitational-wave physicists detailed knowledge of how the merger of any particular pair of masses will produce a particular set of gravitational-wave patterns as they spiral toward each other, approaching one another closer and closer before they suddenly merge into a single object. The timing of the waves that reach each detector, shrinking and expanding each leg in alternation, reveals the masses of the merging objects to a high degree of accuracy. In addition, the detailed time patterns of the events imply that the mergers must involve highly condensed objects, either black holes or neutron stars, and reveal the masses of the merging objects.

The network of gravitational-wave observatories has now detected almost 100 events, each of which involves two merging objects whose masses can be deduced. The great majority of these objects have masses a few dozen times the Sun's. This implies that the mergers arise not only from neutron stars, which have a theoretical mass limit of about 2.2 times the Sun's mass, but also from black holes, which face no such limit. In fact, astrophysicists calculate that typical black holes should have ten to fifty times the Sun's mass, because they form when extremely massive stars collapse by failing to support themselves against their self-gravitation. The data now demonstrate that black hole mergers occurred more often during bygone eras than recently. The greatest merger rate, about 8 billion years ago, coincides with the peak in the star-formation rate. This includes the creation of the most massive stars, which pass through their lifetime in only a few million years and whose death provides the natural source of the most massive black holes.

A few of the gravitational-wave events involve merging objects with masses less than 2.2 times the Sun's, implying that they do arise from neutron stars rather than black holes. In August 2019, the detectors recorded an unusual merger, of one object with 23 times the Sun's mass and another whose mass equals only 2.6 times the Sun's. The former, presumably a typical black hole, poses no problem, but the latter represents either the least massive black hole yet detected, or the most massive neutron star, whose mass must exceed the calculated mass limit. Unless, of course, these calculations turn out to be wrong, or fail to take certain complexities into account. From this single event, we may once again draw the lesson that new techniques for observing the universe almost invariably reveal new aspects of the cosmos. It remains possible that this event, studied through a phenomenon never directly observed until a few years ago, involves a type of object previously unknown to the astronomical community.

From detailed calculations based on Einstein's theory of general relativity, scientists can use the time sequence of gravitational radiation to deduce both the masses of two merging objects and the strength of the gravitational waves that the merger created. Comparison of the detected gravitational-wave strength with the deduced strength at the source of emission provides the distance to the objects. Following this logic, in 2017 the gravitational-wave masters detected waves from two merging neutron stars, which produced what astrophysicists now call a "standard siren": an object whose distance can be inferred from its gravitational waves and whose speed of recession can be found by analyzing the gamma rays that also emerge from the merger. Combining the event's distance with its recession velocity provides a new way to determine the Hubble constant. (The coincidence in arrival time for the gamma rays and gravitational waves also proves that gravitational radiation travels at the speed of light.) The single standard siren so far observed has yielded a Hubble constant value close to that determined with more familiar methods, but with a much larger uncertainty, close to 15 percent. As time goes on, observations of other mergers will improve this accuracy, and will therefore help to resolve the cosmic tension described in Chapter 6.

Carl Sagan liked to say that you had to be made from wood not to stand in awe of what the cosmos has done. Thanks to our improved observations, we now know more than Sagan did about the amazing sequence of events that led to our existence: the quantum fluctuations in the distribution of matter and energy on a scale smaller than the size of a proton that spawned superclusters of galaxies, 30 million light-years across. From chaos to cosmos, this cause-and-effect relationship crosses more than 38 powers of ten in size and 42 powers of ten in time. Like the microscopic strands of DNA that predetermine the identity of a macroscopic species and the unique properties of its members, the modern look and feel of the cosmos was writ in the fabric of its earliest moments, and carried relentlessly through time and space. We feel it when we look up. We feel it when we look down. We feel it when we look within.

PART III
THE ORIGIN OF STARS AND PLANETS

10

DUST TO DUST

If you look at the clear night sky far from city lights, you can swiftly locate a cloudy band of pale light, broken in places by dark splotches, that runs from horizon to horizon. Long known as the (lowercase) "milky way" in the sky, this milk-white haze combines the light from a staggering number of stars and gaseous nebulae. Those who observe the milky way with binoculars or a backyard telescope will see the dark and boring areas resolve themselves into, well, dark and boring areas—but the bright areas will turn from a diffuse glow into countless stars and nebulae.

In his small book *Sidereus nuncius* (*The Starry Messenger*), published in Venice in 1610, Galileo Galilei provided the first account of the heavens as seen through a telescope, including a description of the Milky Way's patches of light. Referring to his instrument as a spyglass, since the name "telescope" ("far-seer" in Greek) had yet to be coined, Galileo could barely contain himself:

The milky way itself, which, with the aid of the spyglass, may be observed so well that all the disputes that for so

many generations have vexed philosophers are destroyed
by visible certainty, and we are liberated from wordy arguments. For the Galaxy is nothing else than a congeries of
innumerable stars distributed in clusters. To whatever
region of it you direct your spyglass, an immense number
of stars immediately offer themselves to view, of which
very many appear rather large and very conspicuous but
the multitude of small ones is truly unfathomable.*

Surely Galileo's "immense number of stars," which delineate
the most densely packed regions of our Milky Way galaxy,
must locate the real astronomical action. Why, then, should
anybody be interested in the intervening dark areas with no
visible stars? Based on their visual appearance, the dark areas
are probably cosmic holes, openings to the infinite and empty
spaces beyond.

Three centuries would pass before anyone figured out that
the dark patches in the milky way on the sky, far from being
holes, actually consist of dense clouds of gas and dust that
obscure more distant star fields and hold stellar nurseries deep
within themselves. Following earlier suggestions by the American astrophysicist George Cary Comstock, who wondered why
faraway stars are much dimmer than their distances alone
would indicate, the Dutch astrophysicist Jacobus Cornelius
Kapteyn in 1909 identified the culprit. In two research papers,
both titled "On the Absorption of Light in Space,"† Kapteyn presented evidence that the dark clouds—his newfound "interstel-

---

* Galileo Galilei, *Sidereus nuncius*, trans. Albert van Helden (Chicago: University of Chicago Press, 1989), 62.
† J. C. Kapteyn, *Astrophysical Journal* 29 (1909): 46; and 30 (1909): 284.

lar medium"—not only block the light from stars but also do so unevenly across the rainbow of colors in a star's spectrum: they absorb and scatter, and therefore attenuate, light at the violet end of the visible spectrum more effectively than they act on red light. This selective absorption preferentially removes more violet than red light, making faraway stars appear redder than nearby ones. The amount of this interstellar reddening of starlight increases in proportion to the total amount of material that the light encounters on its journey to us.

Ordinary hydrogen and helium, the principal constituents of cosmic gas clouds, don't redden light. But molecules made of many other types of atoms do so—especially those that contain the elements carbon and silicon. When interstellar particles grow too large to be called molecules, with hundreds of thousands or millions of individual atoms in each of them, we call them dust. Most of us know dust of the household variety, although few of us care to learn that, in a closed home, dust consists mostly of dead, sloughed-off human skin cells (plus pet dander, if you have one or more live-in mammals). As far as we know, cosmic dust contains nobody's epidermis. However, interstellar dust does include a remarkable ensemble of complex molecules, which emit photons primarily in the infrared and microwave regions of the spectrum. Astrophysicists lacked good microwave telescopes until the 1960s, and effective infrared telescopes until the 1970s. Once they had created these observational instruments, they could investigate the true chemical richness of the stuff that lies between the stars.

During the decades that followed these technological advances, a fascinating, intricate picture of star birth emerged. Not all gas clouds will form stars at all times. More often than not, a cloud finds itself confused about what to do next. Actually, astrophysicists are the confused ones here. We know that an interstellar cloud "wants" to collapse under its own gravity to make one or more stars. But the cloud's rotation, as well as the effects of turbulent gas motions within the cloud, oppose that result. So, too, does the gas pressure that you learned about in high school chemistry class. Magnetic fields can also fight collapse. They penetrate the cloud and constrain the motions of any free-roaming charged particles contained therein, resisting compression and thus impeding the ways in which the cloud can respond to its own gravity. The scary part of this thought exercise comes from the realization that if no one knew in advance that stars existed, frontline research would offer plenty of convincing reasons why stars could never form.

Like the several hundred billion stars in our Milky Way galaxy, giant clouds of gas orbit our galaxy's center. The stars amount to tiny specks, only a few light-seconds across, that float in a vast ocean of nearly empty space, occasionally passing close by one another like ships in the night. Gas clouds, on the other hand, are huge. Typically spanning hundreds of light-years, they each contain as much mass as a million Suns. As these giant clouds lumber through the galaxy, they often collide with one another, entangling their gas- and dust-laden innards. Sometimes, depending on their relative speeds and their angles of impact, the clouds stick together; at other times, adding injury to the insult of collision, they rip each other apart.

If a cloud cools to a sufficiently low temperature (less than about 100 degrees above absolute zero), its constituent atoms will stick together when they collide, rather than careening off one another as they do at higher temperatures. This chemical transition has consequences for everybody. The growing particles—now containing tens of atoms each—begin to scatter visible light to and fro, strongly attenuating the light of the stars behind the cloud. By the time that the particles become full-grown dust grains, they each contain billions of atoms. Aging stars manufacture similar dust grains and blow them gently into interstellar space during their "red giant" phases. Unlike smaller particles, dust grains with billions of atoms no longer scatter the visible light photons from the stars behind them; instead, they absorb those photons and then reradiate their energy as infrared, which can easily escape from the cloud. As this occurs, the pressure from the photons, transmitted to the molecules that absorb it, pushes the cloud in the direction opposite to the direction of the light source. The cloud has now coupled itself to starlight.

Star birth occurs when the forces that make a cloud progressively denser eventually lead to its gravitationally induced collapse, during which each part of the cloud pulls all the other parts much closer. Since hot gas resists compression and collapse more effectively than cool gas does, we face an odd situation. We must cool the cloud before it can ever heat itself by producing a star. In other words, the creation of a star that possesses a 10-million-degree core, sufficiently hot for thermonuclear fusion to begin, requires that the cloud first achieve its coldest possible internal conditions. Only at extremely cold temperatures, a few dozen degrees above absolute zero, can the cloud collapse and allow star formation to begin in earnest.

What happens within a cloud to turn its collapse into newborn stars? Astrophysicists can only gesticulate. Much as they would like to track the internal dynamics of a large, massive interstellar cloud, the creation of a computer model that includes the laws of physics, all the internal and external influences on the cloud, and all the relevant chemical reactions that can occur within it still lies beyond our abilities. A further challenge resides in the humbling fact that the original cloud has a size billions of times larger than that of the star we are trying to create—which in turn has a density 100 sextillion times the average density within the cloud. In these situations, what matters most on one scale of sizes may not be the right thing to worry about on another.

Nevertheless, relying on what we see throughout the cosmos, we can safely assert that within the deepest, darkest, densest regions of an interstellar cloud, where temperatures fall to about 10 degrees above absolute zero, gravity does cause pockets of gas to collapse, easily overcoming the resistance offered by magnetic fields and other impediments. The contraction converts the cloud pocket's gravitational energy into heat. The temperature within each of these regions—soon to become the core of a newborn star—rises rapidly during the collapse, breaking apart all the dust grains in the immediate vicinity as they collide. Eventually, the temperature in the central region of the collapsing gas pocket reaches the crucial value of 10 million degrees on the absolute scale.

At this magic temperature, some of the protons, which are simply naked hydrogen atoms, shorn of the electron that orbits them, move rapidly enough to overcome their mutual repulsion. Their high speeds allow the protons to approach one another closely enough for the strong nuclear force to make them bond. This force, which operates only at extremely short distances, binds together the protons and neutrons in all nuclei. The thermonuclear fusion of protons—"thermo-" because it occurs at high temperatures, and "nuclear fusion" because it fuses particles into a single nucleus—creates helium nuclei, each of which has a mass slightly less than the sum of the particles from which it fused. The mass that disappears during this fusion turns into energy, in a balance described by Einstein's famous equation. The energy embodied in mass (always in an amount equal to the mass times the square of the speed of light) can be converted into other forms of energy, such as additional kinetic energy (energy of motion) of the fast-moving particles that emerge from nuclear fusion reactions.

As the new energy produced by nuclear fusion diffuses outward, the gas heats and glows. Then, at the star's surface, the energy formerly locked in individual nuclei escapes into space in the form of photons, generated by the gas as the energy released through fusion heats it to thousands of degrees. Even though this region of hot gas still resides within the cosmic womb of a giant interstellar cloud, we may nonetheless announce to the Milky Way that . . . a star is born.

Astrophysicists know that stars range in mass from a mere one-tenth of the Sun's to nearly a hundred times our star's mass. For reasons not well understood, a typical giant gas cloud can develop a multitude of cold pockets that all tend to collapse at about the same time to give birth to stars—some puny and others giant. But the odds favor the puny: for every high-mass star, a thousand low-mass stars are born. The fact that no more than a few percent of all the gas in the original cloud participates in star birth presents a classic challenge in explaining star formation: What makes the star-forming tail wag the largely unchanged dog of an interstellar gas cloud? The answer probably lies in the radiation produced by newborn stars, which tends to inhibit further star formation.

We can easily explain the lower bound on the masses of newborn stars. Pockets of collapsing gas with masses less than about one-tenth of the Sun's have too little gravitational energy to raise their core temperatures to the 10 million degrees required for the nuclear fusion of hydrogen. In that case, no nuclear-fusing star will be born; instead, we obtain a failed, would-be star—an object that astrophysicists call a "brown dwarf." With no energy source of its own, a brown dwarf fades steadily, shining from the modest heat generated during the original collapse. The gaseous outer layers of a brown dwarf are so cool that many of the large molecules normally destroyed in the atmospheres of hotter stars remain alive and well within them. Their feeble luminosities make brown dwarfs immensely difficult to detect, so to find them, astrophysicists must employ complex methods similar to those they occasionally use to detect faraway planets: searching for the faint infrared glow from these objects. Only in recent years have astrophysicists discovered brown dwarfs in numbers sufficient to classify them into more than one category.

We can also easily determine the upper mass limit to star formation. A star with a mass greater than about a hundred times the Sun's will have a luminosity so great—such an enormous outpouring of energy in the form of visible light, infrared, and ultraviolet—that any additional gas and dust attracted toward the star will be pushed away by the intense pressure of starlight. The star's photons push on the dust grains within the cloud, which in turn carry the gas away with them. Here starlight couples irreversibly to dust. This radiation pressure operates so effectively that just a few high-mass stars within a dark, obscuring cloud will have luminosities sufficient to disperse nearly all its interstellar matter, laying bare to the universe dozens, if not hundreds, of brand-new stars—all siblings, really—for the rest of the galaxy to see.

Whenever you gaze at the Orion nebula, you can see a stellar nursery of just this sort, located just below the three bright stars that delineate Orion's Belt, at the midpoint of the hunter's somewhat fainter sword. Thousands of stars have been born within this nebula, while thousands more await their birth, soon to create a giant star cluster that will become progressively more visible to the cosmos as the nebula dissipates. The most massive new stars, forming a group called the Orion Trapezium, are busy blowing a giant hole in the middle of the cloud from which they formed. Hubble Telescope images of this region reveal hundreds of new stars in this zone alone, each infant swaddled within a nascent protoplanetary disk made of dust and other molecules drawn from the original cloud. And within each of these disks, a planetary system is forming.

Ten billion years after the Milky Way formed, star formation continues today at multiple locations in our galaxy. Even though most of the star formation that will ever occur in a typical giant galaxy like ours has already taken place, we are fortunate that new stars continue to form, and will do so for many billion years to come. Our good fortune lies in our ability to study the formation process and the youngest stars, seeking clues that will reveal, in all its glory, the complete story of how stars pass from cold gas and dust to luminous maturity.

How old are the stars? No star wears its age on its sleeve, but many show their ages in their spectra. Among the various means that astrophysicists have devised to judge the ages of stars, spectra offer the most reliable hinge for analyzing the different colors of starlight in detail. Every color—every wavelength and frequency of the light waves we observe—tells a story about how matter made the starlight, or affected that light as it left the star, or happened to lie along the line of sight between ourselves and the star. Through close comparison with laboratory spectra, physicists have determined the multitude of ways that different types of atoms and molecules affect the rainbow of colors in visible light. They can apply this fertile knowledge to observations of stellar spectra, and deduce the numbers of atoms and molecules that have affected light from a particular star, as well as the temperature, pressure, and density of those particles. From years of comparing laboratory spectra with the spectra of stars, together with laboratory studies of the spectra of different atoms and molecules, astrophysicists have learned how to read an object's spectrum like a cosmic fingerprint, one that reveals what physical conditions exist within a star's outer layers, the region from which light streams directly outward into space. In addition, astrophysicists can determine how atoms and molecules floating in interstellar space at much cooler temperatures may have affected the spectrum of the starlight they observe, and can likewise deduce the chemical composition, temperature, density, and pressure of this interstellar matter.

In this spectral analysis, each different type of atom or molecule has its own story to tell. The presence of molecules of any type, for example, revealed by their characteristic effects on certain colors in the spectrum, demonstrates that the temperature in a star's outer layers must be less than about 3,000 degrees Celsius (about 5,000° Fahrenheit). At higher temperatures, molecules move so rapidly that their collisions break them apart into individual atoms. By extending this type of analysis over many different substances, astrophysicists can derive a nearly complete picture of the detailed conditions in stellar atmospheres. Some hardworking astrophysicists are said to know far more about the spectra of stars they love than they do about their own families. This may have its downside for interpersonal relations even as it increases human understanding of the cosmos.

Of all nature's elements—of all the different types of atoms that can create patterns in a star's spectrum—astrophysicists recognize and use one in particular to find the ages of the youngest stars. That element is lithium, the third simplest in the periodic table, and familiar to some on Earth as the active ingredient of some antidepressant medications. In the periodic table of the elements, lithium occupies the position immediately after hydrogen and helium, which are deservedly far more famous because they exist in immensely greater amounts throughout the cosmos. During its first few minutes, the universe fused hydrogen into helium nuclei in great numbers, but made only relatively tiny amounts of any heavier nucleus. As a result, lithium remained a rather rare element, distinguished among astrophysicists by the cosmic fact that stars hardly ever make more lithium, but only destroy it. Lithium rides down a one-way street because every star has more effective nuclear fusion reactions to destroy lithium than to create it. As a result, the cosmic supply of lithium has steadily decreased and continues to do so. If you want some, now would be a good time to acquire it.

For astrophysicists, this simple fact about lithium makes it a highly useful tool for measuring the ages of stars. All stars begin their lives with their fair and proportionate share of lithium, left behind by the nuclear fusion that occurred during the universe's first half hour—and during the big bang itself. And what is that fair share? About one in every 100 billion nuclei. After a newborn star begins its life with this "richness" of lithium, things go downhill, lithiumwise, as nuclear reactions within the star's core slowly consume lithium nuclei. The steady and sometimes episodic mixing of matter in the core with matter outside carries material outward, so that after thousands of years, the star's outer layers can reflect what previously happened in its core.

When astrophysicists look for the youngest stars, they therefore follow a simple rule: look for the stars with the greatest abundance of lithium. Each star's number of lithium nuclei in proportion to, for example, hydrogen (determined from careful study of the star's spectrum), will locate the star at some point along a graph that shows how stars' ages correlate with lithium in their outer layers. By using this method, astrophysicists can identify, with confidence, the youngest stars in a cluster, and can assign each of those stars a lithium-based age. Because stars are efficient destroyers of lithium, older stars show little if any of the stuff. Hence the method works well only for stars less than a few hundred million years old. But for these younger stars, the lithium approach works wonders. A recent study of two dozen young stars in the Orion nebula, all of which have masses close to the Sun's, shows ages that range between 1 and 10 million years. Someday astrophysicists may well identify still younger stars, but for now, 1 million years represents about the best they can do.

––––––––

Except for dispersing the cocoons of gas from which they formed, groups of newborn stars bother nobody for a long time, as they quietly fuse hydrogen into helium in their cores and destroy their lithium nuclei as part of their fusion reactions. But nothing lasts forever. Over many million years, in response to the continual gravitational perturbations from enormous clouds that pass by, most would-be star clusters "evaporate," as their members scatter into the general pool of stars in the galaxy.

Nearly 5 billion years after our star formed, the identity of the Sun's siblings has vanished, whether or not those stars remain alive. Of all the stars in the Milky Way and other galaxies, those with low masses consume their fuel so slowly that they live practically forever. Intermediate-mass stars such as our Sun eventually turn into red giants, expanding their outer gas layers a hundredfold in size as they slide toward death. These outer layers become so tenuously connected to the star that they drift into space, exposing a core of spent nuclear fuel that powered the stars' ten-billion-year lives. The gas that returns to space will be swept up by passing clouds, to participate in later rounds of star formation.

Despite their rarity, the highest-mass stars hold nearly all the evolutionary cards. Their high masses give them the greatest stellar luminosities—some of them can boast a million times the Sun's—and because they consume their nuclear fuel far more rapidly than low-mass stars do, they have the shortest lives of all stars, only a few million years, or even less. Continued thermonuclear fusion within high-mass stars allows them to manufacture dozens of elements in their cores, starting with hydrogen and proceeding to helium, carbon, nitrogen, oxygen, neon, magnesium, silicon, calcium, and so on, all the way to iron. These stars forge still more elements in their final fires, which can briefly outshine a star's entire home galaxy. Astrophysicists call each of these outbursts a supernova, similar in appearance (though quite different in their origin) to the Type Ia supernovae described in Chapter 5.

A supernova's explosive energy spreads both the previously made and the freshly minted elements through the galaxy, blowing holes in its distribution of gas and enriching nearby clouds with the raw materials to make new dust grains. The blast moves supersonically through these interstellar clouds, compressing their gas and dust, possibly creating some of the high-density pockets needed to form stars.

The greatest gift to the cosmos from these supernovae appears in many of the elements other than hydrogen and helium—elements such as oxygen, sodium, calcium, aluminum, silicon, iron, and copper, which are capable of forming planets and protists and people. We on Earth live on the product of countless stars that exploded billions of years ago, in epochs of Milky Way history long before our Sun and its planets condensed within the dark and dusty recesses of an interstellar cloud—itself endowed with chemical enrichment furnished from previous generations of high-mass stars.

How did we come to taste this delicious kernel of knowledge, the fact that many of the elements beyond helium were forged within stars? The authors' award for the most underappreciated scientific discovery of the twentieth century goes to the recognition that supernovae—the explosive death throes of high-mass stars—provide a primary source for the origin and abundances of heavy elements in the universe. This relatively unsung realization appeared in a lengthy research article, written by E. Margaret Burbidge, Geoffrey R. Burbidge, William Fowler, and Fred Hoyle, and published in 1957 in the U.S. journal *Reviews of Modern Physics* under the title "Synthesis of the Elements in Stars." In this paper, the four scientists created a theoretical and computational framework that freshly interpreted and melded together four decades of musings by other scientists on two key topics: the sources of stellar energy and the transmutation of chemical elements.

Cosmic nuclear chemistry, the quest to understand how nuclear fusion makes and destroys different types of nuclei, has always been a messy business. The crucial questions have always included: How do the different elements behave when various temperatures and pressures act upon them? Do the elements fuse or do they split? How easily do they do this? Do these processes liberate new kinetic energy or absorb existing kinetic energy? And how do the processes differ for each element in the periodic table?

What does the periodic table of the elements mean to you? If you are like most former students, you will remember a giant chart on the wall of your science classroom, tricked out with mysterious boxes in which cryptic letters and symbols murmured tales of dusty laboratories to be avoided by young souls in transition. But to those who know its secrets, this chart tells a hundred stories of cosmic violence that brought its components into existence. The periodic table lists every known element in the universe, arranged by the increasing number of protons in each nucleus of that element. The two lightest elements are hydrogen, with one proton per nucleus, and helium, with two. As the four authors of the 1957 paper saw, under the right conditions of temperature, density, and pressure, a star can use hydrogen and helium to create all the other elements in the periodic table.

The details of this creation process, and of other interactions that destroy nuclei rather than create them, provide the subject matter for nuclear chemistry, which involves the calculation and use of "collision cross sections" to measure how closely one particle must approach another before they are likely to interact significantly. Physicists can easily calculate collision cross sections for cement mixers, or double-wide mobile homes moving down the street on flatbed trucks, but they face greater challenges in analyzing the behavior of tiny, elusive subatomic particles. A detailed understanding of collision cross sections enables physicists to predict nuclear reaction rates and pathways. Often small uncertainties in their tables of cross sections lead them into wildly erroneous conclusions. Their difficulties resemble what would happen if you tried to navigate your way through one city's subway system with another city's subway map as your guide: your basic theory would be correct, but the details could kill you.

Despite their ignorance of accurate collision cross sections, scientists during the first half of the twentieth century had long suspected that if exotic nuclear processes exist anywhere in the universe, the centers of stars seemed likely places to find them. In 1920, the British theoretical astrophysicist Sir Arthur Eddington published a paper entitled "The Internal Constitution of the Stars," in which he argued that the Cavendish Laboratory in England, the leading center for atomic and nuclear physics research, could not be the only place in the universe that managed to change some elements into others:

> But is it possible to admit that such a transmutation is occurring? It is difficult to assert, but perhaps more difficult to deny, that this is going on . . . and what is possible in the Cavendish Laboratory may not be too difficult in the sun. I think that the suspicion has been generally entertained that the stars are the crucibles in which

the lighter atoms which abound in the nebulæ are com-
pounded into more complex elements.

Eddington's paper, which he expanded into a book of the
same title in 1926, foreshadowed the detailed research of Burbidge, Burbidge, Fowler, and Hoyle, and appeared several
years before the discovery of quantum mechanics, without
which our understanding of the physics of atoms and nuclei
must be judged feeble at best. With remarkable prescience,
Eddington began to formulate a scenario for star-generated
energy via the thermonuclear fusion of hydrogen to helium
and beyond:

> We need not bind ourselves to the formation of helium
> from hydrogen as the sole reaction which supplies the
> energy [to a star], although it would seem that the fur-
> ther stages in building up the elements involve much less
> liberation, and sometimes even absorption, of energy. The
> position may be summarised in these terms: the atoms of
> all elements are built of hydrogen atoms bound together,
> and presumably have at one time been formed from
> hydrogen; the interior of a star seems as likely a place as
> any for the evolution to have occurred.

Any model of the transmutation of the elements ought to
explain the observed mix of elements found on Earth and elsewhere in the universe. To do this, physicists needed to find
the fundamental process with which stars generate energy by
turning one element into another. By 1931, with theories of
quantum mechanics rather well developed (although the neutron had not yet been discovered), the British astrophysicist

Robert d'Escourt Atkinson published an extensive paper, summarized as a "synthesis theory of stellar energy and of the origin of the elements . . . in which the various chemical elements are built up step by step from the lighter ones in stellar interiors, by the successive incorporation of protons and electrons one at a time."

In the same year, the American nuclear chemist William D. Harkins published a paper noting that "elements of low atomic weight [the number of protons plus neutrons in each nucleus] are more abundant than those of high atomic weight and that, on the average, the elements with even atomic numbers [the numbers of protons in each atomic nucleus] are about 10 times more abundant than those with odd atomic numbers of similar value." Harkins surmised that the relative abundances of the elements depend on nuclear fusion rather than on chemical processes such as combustion, and that the heavy elements must have been synthesized from the light ones.

The detailed mechanism of nuclear fusion in stars could ultimately explain the cosmic presence of many elements, especially those that you will obtain each time you add the two-proton, two-neutron helium nucleus to your previously forged element. These constitute the abundant elements with "even atomic numbers" that Harkins described. But the existence and relative numbers of many other elements remained unexplained. Some other means of element buildup must have been at work in the cosmos.

The neutron, discovered in 1932 by the British physicist James Chadwick and notable among elementary particles for possessing zero electric charge, plays a significant role in nuclear fusion that Eddington could not have imagined. To assemble protons requires hard work, because protons naturally repel one another, as do all particles with the same sign of electric charge. To fuse protons, you must bring them sufficiently close (often by way of high temperatures, pressures, and densities) to overcome their mutual repulsion for the strong nuclear force to bind them together. The chargeless neutron, however, repels no other particle, so it can simply march into somebody else's nucleus and join the other assembled particles, held there by the same force that binds the protons. This step does not create another element, which is defined by a different number of *protons* in each nucleus. By adding a neutron, we make an "isotope" of the nucleus of the original element, which differs only in detail from the original nucleus because its total electric charge remains unchanged. For some elements, the freshly captured neutron proves to be unstable once it joins the nucleus. In that case, the neutron spontaneously converts itself into a proton (which stays put in the nucleus) and an electron (which escapes immediately). In this way, like the Greek soldiers who breached the walls of Troy by hiding inside a wooden horse, protons can sneak into a nucleus in the guise of neutrons. If the flow of neutrons stays high, each nucleus can absorb many neutrons before the first one decays. These rapidly absorbed neutrons help to create an ensemble of elements whose origin is identified with the "rapid neutron capture process," and differ from the assortment of elements that result when neutrons are captured slowly, where each successive neutron decays into a proton before the nucleus captures the next one.

Both the rapid and the slow neutron capture processes are responsible for creating many of the elements not otherwise formed through traditional thermonuclear fusion. The remaining elements in nature can be made by a few other processes, including slamming high-energy photons (gamma rays) into the nuclei of heavy atoms, which then break apart into smaller ones.

At the risk of oversimplifying the life cycle of a high-mass star, we may state that each star survives by generating and releasing the energy in its interior that allows the star to support itself against gravity. Without its production of energy through thermonuclear fusion, each stellar ball of gas would simply collapse under its own weight. This fate weighs on stars that exhaust their supplies of hydrogen nuclei (protons) in their cores. After converting its hydrogen into helium, the core of a massive star will next fuse helium into carbon, then carbon to oxygen, oxygen to neon, and so forth up to iron. To successively fuse this sequence of heavier and heavier elements, each one more highly charged because richer in protons, the stellar core must produce successively higher temperatures in order for the nuclei to overcome their natural repulsion. Fortunately, this happens all by itself, because at the end of each intermediate stage, as the star's energy source begins to shut down for lack of fuel, the core's inner regions contract, raising the temperature so that the next pathway of fusion kicks in. Since nothing lasts forever, the star eventually confronts one enormous problem: the fusion of iron does not release energy, but instead absorbs it. This brings bad news to the star, which can now no longer support itself against gravity by pulling a new energy-releasing process out of its nuclear fusion hat. At this point, the star suddenly collapses, forcing its internal temperature to rise so rapidly that a gigantic explosion ensues as the star blows its guts to smithereens.

Throughout each explosion, the availability of neutrons, protons, and energy allows the supernova to create elements in many different ways. In their 1957 article, Burbidge, Burbidge, Fowler, and Hoyle combined (1) the well-tested tenets of quantum mechanics, (2) the physics of explosions, (3) the latest collision cross sections, (4) the varied processes that transmute elements into one another, and (5) the basics of stellar evolutionary theory to implicate supernova explosions decisively as the primary source of the elements heavier than hydrogen and helium in the universe.

With high-mass stars as the source of heavy elements, and supernovae as the smoking gun of element distribution, the fab four acquired the solution to one other problem for free: when you forge elements heavier than hydrogen and helium in stellar cores, you do the rest of the universe no good unless you somehow cast those elements forth into interstellar space, making them available to form worlds with wombats. Burbidge, Burbidge, Fowler, and Hoyle—B^2FH to astrophysicists—unified our understanding of nuclear fusion in stars with the resultant element production visible throughout the universe. Because their conclusions have survived decades of skeptical analysis, their publication stands as a turning point in our knowledge of how the universe works.

In recent years, astrophysicists have realized that stars offer two additional paths to the production of new elements. When high-mass stars become red giants, they expel their outer layers, which form the misleadingly named "planetary nebulae" described in Chapter 8. These gases, rich in carbon and nitrogen produced by nuclear fusion, eventually merge with the material floating through their galaxy and can join a new generation of stars. Most of the carbon and nitrogen essential to life on Earth has followed this path into our ecosystem.

Occupying positions six and seven on the periodic table, car-

bon and nitrogen rank among the lightest elements. The next
30 or so, from oxygen to rubidium, came from the supernova
explosions that B²FH made famous as element factories. Still
heavier elements, astrophysicists now conclude, came from
another arena of cosmic violence: the merger of two neutron
stars, as described in Chapter 9. In 2017, as a key moment in
astronomical progress, three detectors recorded gravitational
radiation from the source known as GW170817. The simultaneous detection of gamma radiation from this object, followed
by studies in visible light and other types of electromagnetic
radiation, created the first example of "multi-messenger
astronomy," the observation of both electromagnetic and gravitational radiation from a single source, an effort that *Science*
magazine named as its "Breakthrough of the Year." Unlike
the merger of two black holes, which generates most of the
gravitational radiation so far detected without leaving significant remnants, neutron-star mergers also produce clouds of
material that emit electromagnetic radiation—gamma rays,
X-rays, visible light, and radio—the details of which reveal the
composition of the debris.

The light from the 2017 event revealed the presence of stron-
tium (element 38), confirming astrophysicists' calculations of
how neutron-star mergers produce heavy elements. These calculations show that a significant portion, and in many cases the
great majority, of the heaviest elements, which arise from the
rapid neutron capture process described above, owe their existence to the tightening spirals and eventual merger of neutron
stars in binary systems. Since neutron stars represent the collapsed cores of former supernovae, supernova enthusiasts could
rightly claim that here, too, we encounter element production
from exploding stars, simply at one remove from the explosion.

Yes, Earth and all its life comes from stardust. No, we have not solved all of our cosmic chemical questions. A curious contemporary mystery involves the element technetium, which, in 1937, became the first element to be created artificially in Earthbound laboratories. (The word "technetium," along with others that use the prefix "tech-," derives from the Greek *technetos*, which translates to "artificial.") We have yet to discover technetium on Earth, but astrophysicists have found it in the atmospheres of a small fraction of the red-giant stars in our galaxy. This would hardly surprise us were it not for the fact that technetium decays to form other elements, and does so with a half-life of a mere 2 million years, far shorter than the age and life expectancy of the stars in which we observe it. This conundrum has led to exotic theories that have only recently achieved consensus within the community of astrophysicists.

To interested scientists, these ongoing chemical mysteries have an allure as strong as the questions related to black holes, quasars, and the early universe. But you hardly ever read about them. Why? Because, quite typically, the media has predetermined what deserves coverage and what does not. Apparently the news about the cosmic origins of every element in your body and your planet doesn't make the cut. And yet, as Harlow Shapley, the director of the Harvard Observatory, mused in his 1963 book *The View from a Distant Star,* "Mankind is made of star stuff, ruled by universal laws."

II

WHEN WORLDS WERE YOUNG

In our attempts to uncover the history of the cosmos, we have continually discovered that the segments most deeply shrouded in mystery are those that deal with *origins*—of the universe itself, of its most massive structures (galaxies and galaxy clusters), of the stars that provide most of the light in the cosmos, and of the planets that surround those stars. Each of these origin stories fills a vital role, not only in explaining how an apparently formless cosmos produced complex assemblages of different types of objects but also in determining how and why, 14 billion years after the big bang, we now find ourselves alive on Earth to solve the basic cosmic mystery, How did this all happen?

This mystery arises in large part because during the cosmic "dark ages," when matter was just beginning to distribute itself into self-contained units such as stars and galaxies, most of this matter generated little or no detectable radiation. The dark ages have left us with only the barest possibilities, still imperfectly explored, for observing matter during its early stages of organization. This in turn implies that we must rely, to an uneasily large extent, on our theories of how matter ought to behave, with relatively few points at which we can check these theories against observational data.

When we turn to the origin of planets, the mysteries deepen. We lack not only *observations* of the crucial, initial stages of planetary formation but also successful *theories* of how the planets began to form. To celebrate the positive, we note that the question "What made the planets?" has grown considerably broader in recent years. Throughout most of the twentieth century, this question centered on the Sun's family of planets. In the twenty-first century, astrophysicists' discovery of thousands of "exoplanets" around relatively nearby stars has provided them with far more data from which to deduce the early history of planets, and in particular to determine how these astronomically small, dark, and dense objects formed along with the stars that give them light and, potentially, life.

Astrophysicists may now have more data, but they have no better answers than before. Indeed, the discovery of exoplanets, many of which move in orbits far different from those of the Sun's planets, has in many ways confused the issue, leaving the story of planet formation no closer to closure. In simple summary, we can state that no good explanation exists of how the planets *began* to build themselves from gas and dust, though we can easily perceive how the formation process, once well underway, made larger objects from smaller ones, and did so within a rather brief span of time.

The beginnings of planet building pose a remarkably intractable problem, to the point that one of the world's experts on the subject, Scott Tremaine of Princeton University, has elucidated (partly in jest) Tremaine's laws of planet formation. The first of these laws states that "all theoretical predictions about the properties of exoplanets are wrong," and the second that "the most secure prediction about planet formation is that it can't happen." Tremaine's humor underscores the ineluctable fact that planets do exist, despite our inability to explain this astronomical enigma.

More than two centuries ago, attempting to explain the formation of the Sun and its planets, Immanuel Kant proposed a "nebular hypothesis," according to which a swirling mass of gas and dust that surrounded our star-in-formation condensed into clumps that became the planets. In its broad outlines, Kant's hypothesis remains the basis for modern astronomical approaches to planet formation, having triumphed over the opposing concept, much in vogue during the first half of the twentieth century, that the Sun's planets arose from a close passage of another star by the Sun. In that scenario, the gravitational forces between the stars would have drawn masses of gas from each of them, and some of this gas could then have cooled and condensed to form the planets. This hypothesis, promoted by the famed British astrophysicist James Jeans, had the defect (or the appeal, for those inclined in that direction) of making planetary systems extremely rare, because sufficiently close encounters between stars probably occur only a few times during the lifetime of an entire galaxy. Once astrophysicists calculated that almost all the gas pulled from the stars would disperse rather than condense, they abandoned Jeans's hypothesis and returned to Kant's, which implies that many, if not most, stars should have planets in orbit around them—a hypothesis richly verified by the flood of newly discovered exoplanets.

Astrophysicists do have good evidence that stars form not one by one but by the thousands and tens of thousands, within giant clouds of gas and dust that may eventually give birth to about a million individual stars. One of these giant stellar nurseries has produced the Orion nebula, the closest large star-forming region to the solar system. Within a few million years, this region will have produced hundreds of thousands of new stars, which will blow most of the nebula's remaining gas and dust into space, so that astrophysicists a hundred thousand generations from now will observe the young stars unencumbered by the remnants of their star-birthing cocoons.

With radio telescopes, astrophysicists can map the distribution of cool gas and dust in the immediate vicinities of young stars. Their maps typically show that young stars do not sail through space devoid of all surrounding matter; instead, the stars usually have orbiting disks of matter, similar in size to the solar system, but made of hydrogen gas (and of other gases in lesser abundances) sprinkled throughout with dust particles. The term "dust" describes groups of particles that each contain several million atoms and have sizes much smaller than that of the period that ends this sentence. Many of these dust grains consist primarily of carbon atoms, linked together to form graphite (the chief constituent of the "lead" in a pencil). Others are mixtures of silicon and oxygen atoms—in essence tiny rocks, with mantles of ice surrounding their stony cores.

The formation of these dust particles in interstellar space has its own mysteries and detailed theories, which we may skip past with the happy thought that the cosmos *is* dusty. To make this dust, atoms have come together by the millions; in view of the extremely low densities between the stars, the likeliest sites for this process seem to be the extended outer atmospheres of cool stars, which gently blow material into space.

The new space telescope's infrared capabilities, which allow it to peer into the long-vanished epoch of galaxy formation, will also aid astrophysicists in their efforts to understand how planets form. Because infrared radiation penetrates dust-rich regions far more readily than visible light does, the JWST can examine star-forming regions, along with their disks of surrounding material that can form planets, in highly improved detail.

———————

The production of interstellar dust particles provides an essential first step on the road to planets. This holds true not only for solid planets like our own but also for gas-giant planets, typified in the Sun's family by Jupiter and Saturn. Even though these planets consist primarily of hydrogen and helium, astrophysicists have concluded from their calculations of the planets' internal structure, along with their measurements of the planets' masses, that the gas giants must have solid cores. Of Jupiter's total mass, 318 times Earth's, several dozen Earth masses reside in a solid core. Saturn, with 95 times Earth's mass, also has a solid core, with one or two dozen times the mass of Earth. The Sun's two smaller gas-giant planets, Uranus and Neptune, have proportionally larger solid cores. In these planets, with 15 and 17 times Earth's mass, respectively, the core may contain more than half of the planet's mass.

For all four of these planets, and presumably for all of the giant planets recently discovered around other stars, the planetary cores played an essential role in the formation process: first came the core, and then came the gas, attracted by the solid core. Thus all planet formation requires that a large lump of solid matter must form first. Of the Sun's planets, Jupiter has the largest of these cores, Saturn the next largest, Neptune the next, Uranus after that, and Earth ranks fifth, just as it does in total size. The formation histories of all the planets pose a fundamental question: How does nature make dust coagulate to form clumps of matter many thousand miles across?

The answer has two parts, one known and one unknown, with the unknown part, not surprisingly, closer to the origin. Once you form objects half a mile across, which astrophysicists call planetesimals, each of them will have sufficiently strong gravity to attract other such objects successfully. The mutual gravitational forces among planetesimals will build first planetary cores and then planets at a brisk pace, so that a few million years will take you from millions of clumps, each the size of a small town, to entire new worlds, ripe to acquire either a thin coat of atmospheric gases (in the case of Venus, Earth, and Mars) or an immensely thick one of hydrogen and helium (for the four gas-giant planets, which orbit the Sun at distances large enough for them to accumulate huge quantities of these two lightest gases). To astrophysicists, the transition from half-mile-wide planetesimals to planets reduces to a series of well-understood computer models that produce a wide variety of planetary details, but almost always yield inner planets that are small, rocky, and dense, as well as outer planets that are large and (except for their cores) gaseous and rarefied. During this process, many of the planetesimals, as well as some of the larger objects that they make, find themselves flung entirely out of the solar system by gravitational interactions with still larger objects.

All this works rather well on a computer, but building the half-mile-wide planetesimals in the first place eludes even our finest astrophysicists' present abilities to integrate their knowledge of physics with their computer programs. Gravity can't make planetesimals, because the modest gravitational forces between small objects won't hold them together effectively. Two theoretical possibilities exist for making planetesimals from dust, neither of them highly satisfactory. One model proposes the formation of planetesimals through accretion, which occurs when dust particles collide and stick together. Accretion works well in principle, because most dust particles *do* stick together when they meet. This explains the origin of dust bunnies under your couch, and if you imagine superdust bunnies growing around the Sun, you can, with only minimal mental effort, let them grow to become beach-ball-sized, house-sized, block-sized, and before long the size of planetesimals, ready for serious gravitational action.

Unfortunately, unlike the production of actual bunnies, the dust-bunny growth of planetesimals seems to require far too much time. Radioactive dating of unstable nuclei detected in the oldest meteorites implies that the formation of the solar system required no more than a few tens of millions of years, and quite possibly a good deal less time than that. In comparison with the current age of the planets, approximately 4.55 billion years, this amounts to a dram in the bucket, only 1 percent (or less) of the total span of the solar system's existence. The accretion process requires significantly longer than a few tens of millions of years to make planetesimals from dust; so unless astrophysicists have missed something important in understanding how dust accumulates to build large structures, we need another mechanism to surmount the time barriers to planetesimal formation.

That other mechanism may consist of giant vortices that sweep up dust particles by the trillions, whirling them quickly toward their happy agglomeration into significantly larger objects. Because the contracting cloud of gas and dust that became the Sun and its planets apparently acquired some rotation, it soon changed its overall shape from spherical to platelike, leaving the Sun-in-formation as a relatively dense contracting sphere at the center, surrounded by a highly flattened disk of material in orbit around that sphere. To this day, the orbits of the Sun's planets, which all follow the same direction and lie in nearly the same plane, testify to a disklike distribution of the matter that built the planetesimals and planets. Astrophysicists envision that within such a rotating disk, rippling "instabilities"—alternating regions of greater and lesser density—will appear. The denser parts of these instabilities collect both gaseous material and dust that floats within the gas. Within a few thousand years, these instabilities become swirling vortices that can sweep large amounts of dust into relatively small volumes.

This vortex model for the formation of planetesimals shows promise, though it has not yet won the hearts of those who seek explanations of how the solar system produced what young planets need. Upon detailed examination, the model provides better explanations for the cores of Jupiter and Saturn than for those of Uranus and Neptune. Because astrophysicists have no way to prove that the instabilities needed for the model to work actually did occur, we must refrain from passing judgment ourselves. The existence of numerous small asteroids and comets, which resemble planetesimals in their sizes and compositions, support the concept that billions of years ago, planetesimals by the millions built the planets. Let us therefore regard the formation of planetesimals as an established, if poorly understood, phenomenon that somehow bridges a key gap in our knowledge, leaving us ready to admire what happens when planetesimals collide.

In this scenario, we can easily imagine that once the gas and dust surrounding the Sun had formed a few trillion planetesimals, this armada of objects collided, built larger objects, and eventually created the Sun's four inner planets and the cores of its four giant planets. We should not overlook the planets' moons, smaller objects that orbit all of the Sun's planets except the innermost, Mercury and Venus. The largest of these moons, with diameters of a few hundred to a few thousand miles, appear to fit nicely into the model that we have created, because they presumably also arose from planetesimal collisions. Moon building ceased once collisions had built the satellite worlds to their present sizes, no doubt (we may assume) because by that time the nearby planets, with their stronger gravity, had taken possession of most of the nearby planetesimals. We should include in this picture the hundreds of thousands of asteroids that orbit between Mars and Jupiter. The largest asteroids, a few hundred miles in diameter (much smaller than the one Bruce Willis encountered in the 1998 film *Armageddon*), should likewise have grown through planetesimal collisions, and should then have found themselves stymied from further growth by gravitational interference from the nearby giant planet Jupiter. The smaller asteroids, less than a mile across, may represent naked planetesimals, objects that grew from dust but never collided with one another, once again thanks to Jupiter's influence, after attaining sizes ripe for gravitational interaction.

For the moons that orbit the giant planets, this scenario seems to work quite well. All four giant planets have families of satellites that range in size from the large or extremely large (up to the size of Mercury) down to the small or even astronomically minuscule. The pygmies among these moons, less than a mile across, may again be naked planetesimals, deprived of any further collisional growth by the presence of nearby objects that had already grown much larger. In each of these four families of satellites, almost all of the larger moons orbit the planet in the same direction and in nearly the same plane. We can hardly refrain from explaining this result with the same cause that made the planets orbit in the same direction and nearly the same plane: around each planet, a rotating cloud of gas and dust produced clumps of matter, which grew to planetesimal and then to moon sizes.

In the inner solar system, only our Earth has a sizable moon. Mercury and Venus have none, while Mars's two potato-shaped moons, Phobos and Deimos, each span only a few miles, and should therefore represent the earliest stages of forming larger objects from planetesimals. Some theories assign the origin of these moons to the asteroid belt, with their present orbits around Mars the result of Mars's gravitational success in capturing these two former asteroids.

And what of our Moon, more than two thousand miles in diameter, surpassed in size only by Titan, Ganymede, Triton, and Callisto (and effectively tied with Io and Europa) among all the moons of the solar system? Did the Moon also grow from planetesimal collisions, as the four inner planets did?

This seemed quite a reasonable supposition until humans brought lunar rocks back to Earth for detailed examination. More than three decades ago, the chemical composition of the rock samples returned by the Apollo missions imposed two conclusions, one on either side of the possibilities for the Moon's origin. On the one hand, the composition of these Moon rocks resembles that of rocks on Earth so closely that the hypothesis that our satellite formed entirely apart from us no longer seems tenable. On the other hand, the Moon's composition differs from Earth's just enough to prove that the Moon did not entirely form from terrestrial material. But if the Moon did not form apart from Earth, and was not made from Earth, how did it form?

The current answer to this conundrum, amazing though it may seem on its surface, builds upon a once-popular hypothesis that the Moon formed as the result of a giant impact that flung large amounts of the Earth's material into space, where some of it coalesced to form our satellite. Under the new view, which has already gained wide acceptance as the best available explanation, the Moon *did* form as the result of a giant object that struck Earth, and the impacting object, which astrophysicists have named Theia after the mother of the Moon goddess Selene, was so large—about the size of Mars—that it naturally added some of its material to the matter ejected from Earth. Much of the material thrown into space by the force of the impact might have vanished from our immediate vicinity, but enough remained behind to coagulate into our familiar Moon, made of Earth plus foreign matter. All of this occurred 4.5 billion years ago, during the first 100 million years after the formation of the planets began.

If a Mars-sized object struck Earth in that bygone era, where is it today? The impact could hardly have knocked the object into pieces so small that we cannot observe them: our finest telescopes can find objects in the inner solar system as small as the planetesimals that built the planets. The answer to this objection takes us to a new picture of the early solar system, one that emphasizes its violent, collisional nature. The fact that planetesimals built a Mars-sized object, for example, did not guarantee that this object would endure for long. Not only did this object collide with Earth, but the good-sized pieces produced by that collision would also have continued to collide with Earth and the other inner planets, with one another, and with the Moon (once it had formed). In other words, collisional terror reigned over the inner solar system during its first several hundred million years, and the pieces of giant objects that struck the planets as they formed themselves became part of these planets. Theia's impact on Earth ranked merely among the largest in a rain of bombardment, an epoch of destruction that brought planetesimals and much larger objects crashing down on Earth and its neighbors.

Seen from another perspective, this death-dealing bombardment simply marked the formation process's final stages. The process culminated in the solar system we see today, little changed during 4 billion years and more: one ordinary star, orbited by eight planets (plus icy Pluto, more akin to a giant comet than to a planet), hundreds of thousands of asteroids, trillions of meteoroids (smaller fragments that strike Earth by the thousands every day), and trillions of comets—dirty snowballs that formed at dozens of times Earth's distance from the Sun. We must not forget the planets' satellites, which have moved, with few exceptions, in orbits with long-term stability ever since their birth, 4.6 billion years ago.

Throughout most of history, our solar system provided us

with the only known planets. Quite naturally, even as they sought to discover planets around other stars, astrophysicists expected that any such planets would generally resemble the Sun's family in their sizes and orbits. How wrong they were! How much more fascinating reality turned out to be than our imagination!

12

PLANETS BEYOND THE SOLAR SYSTEM

Thro' worlds unnumbered tho' the God be known,

'Tis ours to trace him only in our own.

He, who through vast immensity can pierce,

See worlds on worlds compose one universe,

Observe how system into system runs

What other planets circle other suns,

What varied Being peoples ev'ry star,

May tell why Heav'n has made us as we are.

—ALEXANDER POPE, *AN ESSAY ON MAN* (1733)

Nearly five centuries ago, Nicolaus Copernicus resurrected a hypothesis that the ancient Greek astronomer Aristarchus had first suggested. Far from occupying the center of the cosmos, said Copernicus, Earth belongs to the family of planets that orbit the Sun. Even though billions of humans have yet to accept this fact, believing in their hearts that Earth remains immobile as the heavens turn around her, astronomers have long offered convincing arguments that Copernicus wrote the truth about the nature of our cosmic home. The conclusion that Earth ranks as just one of the Sun's planets immediately suggests that other planets fundamentally resemble our own, and that they may well possess their own inhabitants, potentially endowed as we are with plans and dreams, work, play, and fantasy.

Planets around other stars therefore offer us a cosmic laboratory in which we may someday explore when and how life appears in the cosmos, how it evolves, and how it affects the worlds on which it lives. Earth's planetary neighbors in the solar system offer a limited set of examples among Alexander Pope's "worlds unnumbered." Prospecting on Mars, and elsewhere in the solar system, could succeed in finding life on our neighbor worlds. On the other hand, we may find that our planet formed within the band of distances from the Sun that produce temperatures favorable to life, leaving other spheres outside the "habitable zone." Our descendants may someday investigate planetary systems close to our own not only for their intrinsic interest, but also for what they can teach us about life's manifold possibilities throughout the Milky Way and beyond.

Before that happens, we may salute how far we have come within a single human generation. For centuries, astronomers used telescopes to observe hundreds of thousands of individual stars without the ability to discern whether or not any of these stars have planets of their own. Their observations did reveal that our Sun ranks as an entirely representative star, whose near twins exist in great numbers throughout our Milky Way galaxy. If the Sun has a planetary family, so too might other stars, with their planets equally capable of giving life to creatures of all possible forms. Expressing this view in a manner that affronted papal authority brought Giordano Bruno to his death at the stake in 1600. Today, tourists returning to Rome can pick their way through the crowds at the outdoor cafés in the Campo de' Fiori to reach Bruno's statue at its center, then pause for a moment to reflect on the power of ideas (if not the power of those who hold them) to triumph over those who would suppress them.

As Bruno's fate helps to illustrate, imagining life on other worlds ranks among the most powerful ideas ever to enter human minds. Were this not so, Bruno would have lived to a riper age, and NASA would probably find itself shorter of funds. Thus speculation about life on other worlds has focused throughout history, as NASA's attention still does, on the planets that orbit the Sun. In our search for life beyond Earth, however, a great frost has appeared: none of the other worlds in our solar system seem particularly fit for life.

Although this conclusion hardly does justice to the scores of possible paths by which life might arise and maintain itself, the fact remains that our initial explorations of Mars and Venus, as well as of Jupiter, Saturn, and their major moons, have failed to produce any convincing signs of life. To the contrary, we have found a great deal of evidence for conditions extremely hostile to life as we know it, though hardly so extreme as to rule life beyond possibility. Much more searching remains to be done, and fortunately (for those who engage themselves mentally in this effort) continues, especially in the hunt for life on Mars. Nevertheless, the verdict on extraterrestrial life in the solar system shows sufficient likelihood of proving negative that supple minds often look beyond our cosmic neighborhood, to the vast array of possible worlds that orbit stars other than our Sun.

———

Until 1995, speculation about planets around other stars could proceed almost entirely unfettered by facts. With the exception of a few pieces of Earth-sized debris in orbit around the remnants of supernovae, which almost certainly formed after the stars exploded and barely qualify as planets, astrophysicists had never found a single exoplanet, a world orbiting a star other than the Sun. At the end of that year came the dramatic announcement of the first such discovery; then, a few months later, came four more; and then, with the floodgates open, finding new worlds proceeded ever more swiftly. Today, we know of 5,000 exoplanets, hundreds of times more than the familiar worlds that orbit the Sun. This number will grow for years to come, thanks to recently launched satellites devoted to searching for planets around other stars.

Astrophysicists' ingenuity and hard work has led them to develop at least eight different methods to discover exoplanets. Two of these techniques have yielded the great bulk of these discoveries; two more have each revealed more than 150 new planets; while the final four, taken all together, now total more than 100 discoveries. We may content ourselves with an examination of the first four of these eight methods, starting with the two lesser ones and then celebrating the triumphant pair.

First comes direct imaging, the most straightforward method for finding exoplanets by seeing them directly with a powerful optical system. Obvious though it may appear, this approach faces an enormous problem that stymied astrophysicists for decades: astronomically speaking, planets nestle right next to their stars, and shine only weakly by reflecting starlight. A distant observer of our solar system, ready to spot Jupiter, must contend with the fact that the Sun outshines its largest planet by a factor of a billion. Observations made in infrared radiation rather than visible light can help: the brightness factor falls to a million, but still poses an enormous problem for resolving a close pair of unequal images. As a result, the 150 or so exoplanets glimpsed directly share two common characteristics: they are as large as, or larger than, Jupiter, and most of them orbit their stars at distances ranging between three and a hundred times Saturn's distance from the Sun. Even so, each hard-won image of one of these exoplanets typically appears as a fuzzy blob, incapable of exciting our visual channels of attention. Aside from these victories, astrophysicists have never *seen* any of the rest of the 5,000 known individual exoplanets in 3,600 planetary systems.

What you might regard as a serious shortcoming actually represents a triumph of how science works around such challenges. The second planet-finding procedure has the name "gravitational lensing," which once again brings Albert Einstein's insights into play. Einstein's general theory of relativity tells us that gravitational forces bend space, and therefore bend the paths of light rays that pass close by a massive object such as a star. If a star's motion through space happens to carry it in front of a more distant star, the closer star's gravity will focus the distant star's light as a sort of lens, producing a sharp spike in the distant star's observed brightness. If the closer star has one or more planets, they will each produce a similar, though much briefer and far less pronounced, increase in brightness. The secondary spikes' sizes depend on the masses of the objects that produce them, while the exact timing between the primary and secondary effects depends on the star-planet distances. Surveying large numbers of stars on every clear night, obtaining ever-more-precise measurements of their brightnesses, astrophysicists have used telescopes in Australia and the United States to discover well over 150 exoplanets, close to the number found by direct imaging. This technique works well for planetary systems with distances from us greater than those searchable by competing methods, but its discoveries are all one-offs, since a star's motion will never return it to the same position with respect to its more distant neighbors.

The two major techniques for discovering exoplanets rely on observations of stars themselves rather than their planets. Astrophysicists' direct observation of stars can reveal either a brief, modest reduction in the stars' *brightnesses*, or periodic, repetitive changes in the stars' *motions* through space (the Doppler-effect method). From careful analysis of these changes in a star's brightness or motions, astrophysicists can deduce the existence of one or more planets around the star, and can determine a surprising range of the planets' characteristics.

"Transit," a sweet and ancient astronomical term, refers to the passage of one object directly in front of another (a purist might therefore insist that gravitational-lensing events should also be classified as transits). A "transit of Venus" occurs when Venus comes directly between Earth and the Sun, as it soon will again (in 2117 and 2125). An exoplanet will transit its star if, and only if, the plane of its orbit around its star aligns with our line of sight to the star. If this alignment does not occur, we lose the chance to discover the star's planets by their transits, but in the small fraction of situations where it does, the method gives fine fruit.

To make it work, astrophysicists must first find the dip in brightness that signals a transit, then observe several successive transits, and verify that the time intervals between them remain constant, in order to reject the possibility that they have found an anomaly of the star itself. This verification by time interval immediately reveals a planet's orbital period, and the amount of the starlight droop shows the planet's size. Jupiter, for instance, would reduce the Sun's light by 1 percent when seen in transit every 12 years, while Earth would do so by 0.01 percent every year. Our atmosphere's ever-waving motion, which makes stars twinkle as we see them, eliminates the possibility of using ground-based observatories to make the precise measurements that astrophysicists seek, but satellites avoid this difficulty and can find exoplanets even smaller than Earth. Planets with shorter orbital periods reveal themselves more rapidly, while those whose periods span years naturally require longer periods of observation to verify their existence.

The transit method has taken first place among exoplanet-discovery techniques by more than a factor of three, with 3,500 planet discoveries—for now. Building on the impressive results of NASA's Kepler satellite (launched in 2009 and deactivated in 2018), the currently operating transit hunters—NASA's TESS (Transiting Exoplanet Survey Satellite, launched in 2018) and the European Space Agency's CHEOPS (CHaracterising ExOPlanets Satellite, launched in 2019)—will be joined later by ESA's PLATO (PLAnetary Transits and Oscillations of stars) satellite as astrophysicists steadily improve the numbers and deduce the properties of exoplanets found by their transits. The transit instruments either survey a large number of stars (150,000 for Kepler, 200,000 for TESS) to discover exoplanets, or study in detail exoplanets found by other techniques, as CHEOPS does.

The other highly successful detection method, with close to a thousand notches on its exoplanet belt, relies on the Doppler effect, which we met in Chapter 5. It describes how the light from a galaxy—the combination of light from billions of stars—can reveal the galaxy's motion toward us or away from us. When astrophysicists apply the same analysis to the light from stars in the Milky Way, they can measure the velocity with which individual stars approach or recede from us; it does not matter whether we or the star (or both) are in motion. For stars that circle the Milky Way without planets, this velocity should remain constant over human time scales, but if one or more planets accompany a star, their gravitational forces, though comparatively weak, will pull the star a bit, first in one direction and then in another, as the planets proceed along their orbits. From this simple fact, an inevitable consequence of Newton's laws of motion and gravitation, have sprung empires of knowledge about a host of exoplanets.

If astrophysicists find that a star's velocity along our line of sight rises a bit above its average value, passes through that average, falls below it, rises again, and continues to repeat this cycle of variation, they conclude that the velocity changes indeed arise from a planet moving in orbit around the star, tugging it first a bit toward us and then a bit away from us. If so, then the length of each cycle of variation equals the length of time for the planet to orbit the star. Still more information flows from astrophysicists' knowledge of the stars that these planets orbit. From their detailed studies of stellar spectra, astrophysicists can assign each star its rate of energy output, from the dimmest, least massive stars, which possess only about one-tenth of the Sun's mass, to the most luminous, most massive examples, with many dozen times the mass of the Sun. Once they know the star's mass as well as the planet's orbital period, they can deduce the average planet-star distance. At any particular distance, a more massive star will require a more rapid motion to keep the planet in orbit, and for any star, more distant planets will orbit more slowly, as is true for the Sun's flock.

Astrophysicists can derive still more from a graph of the changes induced by the Doppler effect. This graph reveals the *shapes* of the planet's and the star's orbits around their common center of mass. These orbital shapes are identical, even though the sizes of the orbits vary in inverse proportion to the objects' masses. Circular orbits produce an even, sinusoidal variation, while elongated orbits skew the peaks and valleys toward one side or the other, and do so more for greater amounts of elongation.

But even more significant information resides in the data that describe the Doppler dance of the star and its planet. Armed with their knowledge of the star's mass, astrophysicists can infer the mass of a planet that orbits at any given distance. The star's velocity changes arise from the planet's gravitational force upon it, which depends on the planet's mass and its distance from its star. With the distance known, the planet's mass emerges quickly, with a disclaimer. Astrophysicists typically have no way to determine whether they are looking at the planet's orbit edge-on, or—more likely—from a tilted angle, so that the planet rises "above" the line of sight to the star for half of its orbit and falls "below" that line for the other half. In that case, astrophysicists see only part of the planet's full effect on the star's motion. As a result, the planetary masses that they infer from the maximum changes in the star's velocity represent only *minimum* masses, each of which would be the entire mass only if the plane of the planet's orbit happened to lie exactly parallel to our line of sight.

Because we may reasonably expect a random distribution
in the sizes of the angle by which a particular planet's orbital
plane tilts from our line of sight, geometry allows the conclusion that the average planetary mass gleaned from Doppler-
effect observations of stellar velocities equals just half of the
actual average mass. We typically do not know which planets
have more than this, and which less. But in the happy minority of cases for which the planet actually transits its star, we
do know that we observe its entire effect, so the deduced mass
equals the actual mass. These transit-plus-Doppler-effect exoplanets qualify among those for which astrophysicists have the
most complete information.

Armed with the star's energy output and the planet-star
distance, astrophysicists can calculate an exoplanet's surface
temperature, which adds in many cases to their knowledge
of the planet's size, mass, orbital shape, and orbital period.
Furthermore, through many careful measurements, they have
found many cases of more than one planet in orbit around a
distant star. Each planet can have its own transit, or add to
the Doppler-effect changes produced in a star's motion, in ways
that can be statistically disentangled to reveal an entire planetary system. By now, astrophysicists have discovered more
than 800 systems that contain two or more exoplanets. One
of them ties our Sun by possessing eight planets; another has
seven; and six others have six planets each.

When we seek to analyze what our current data that
describe 5,000 exoplanets imply for the search for extraterrestrial life, our understanding of the range among exoplanets'
masses, orbital sizes and periods, and surface temperatures
provides a fine basis for speculation. We cannot, of course,
claim that we have a fully unbiased sampling of the planetary

census of the Milky Way galaxy. Each of the four primary detection methods inevitably introduces a particular bias. Direct imaging succeeds for large planets at comparatively great distances from their stars. Gravitational lensing works best for planets with large masses, without much regard to the planet-star distances. The transit approach favors large planets, somewhat favors planets with shorter orbital periods, and excludes planets with orbital periods longer than a decade or so. Searching for exoplanets through the Doppler effect on stars' observed motions prejudices the results in favor of close-in, high-mass planets.

Fully aware of the biases inherent in each of their search techniques, astrophysicists have reached important generalizations about exoplanets. The most obvious and significant lies in their abundance. Quite possibly, stars with planets outnumber those without any. Even the closest star to the Sun, a dim red dwarf named Proxima Centauri (a distant outlier of the Alpha Centauri double-star system), has its own planet. Next most notably and significantly, planets possess an enormous range in their sizes, masses, orbital periods, and planet-star distances. Planets exist with diameters less than 1/3 of Earth's (and most probably, only our limited abilities prevent us from finding still smaller planets), while other planets have diameters 8 times Jupiter's. Some planets have less than 1/10,000 of Earth's mass, and others have a dozen times Jupiter's 318 Earth masses. Some exoplanets take only 40 minutes to orbit their stars, while others require hundreds of years. Connected to this fact, some have orbits only 1/350 the size of Earth's, while others move in orbits similar to Neptune's, or even larger ones.

Speaking broadly, the most common type of exoplanet found among the first 4,000 (three-quarters of them revealed by their transits) has turned out to differ from *all* of the Sun's planets! Exoplanetary astrophysicists call these "super-Earths" to denote a planet significantly larger than Earth but significantly smaller than any of the Sun's four giants. These super-Earths, most often about twice the size of our planet, typically take only a few months to orbit their stars. In fact, approximately half of all stars have at least one planet with an orbital period no greater than 100 days. (The Sun qualifies, since Mercury takes 88 days for each orbit.) The existence of super-Earths, almost completely unsuspected before the age of exoplanet discovery, serves as an excellent reminder of the danger of drawing conclusions from a limited set of data—in this case, the Sun's eight planets. The larger super-Earths may prove to be mostly gaseous; if so, then Earth may represent one of the largest rocky planets that the formation process has let loose upon the cosmos.

Another result with broad implications for life in the universe concerns the nature of the stars with exoplanets. Many of these are Sunlike stars, just the sort that we tend to imagine in considering life in other planetary systems. But an even larger number of stars known to possess planets—perhaps because their low masses make detecting their planets an easier task—are low-luminosity red-dwarf stars, typically with about one-quarter of the Sun's mass, one-third of its diameter, one-half of its surface temperature, and 1 or 2 percent of its energy output. These red dwarfs dominate our galaxy by numbers, though their low luminosities make them difficult to detect outside of our nearby surroundings. Their low energy output, even in view of their low masses, gives them the longest lifetimes of any stars: trillions of years, rather than the mere 10 billion or so for a Sunlike star. Unsurprisingly, planets around red dwarfs orbit at comparatively tiny distances. Many of these have orbital periods measured in hours, and orbital sizes a small fraction of Mercury's. Crucially in the prospects for life, many of these planets have surface temperatures comparable to Earth's. An orbit close to a dim red star can yield results similar to orbiting a much more luminous star at a much greater distance.

As proud residents of the twenty-first century, we can take delight in possessing the basic facts about planets in the Milky Way. We know, with reasonable accuracy, the distribution of their sizes, their masses, their surface temperatures, and their orbital shapes, sizes, and periods. Many of them qualify as roughly Earthlike planets, in Earthlike orbits around Sunlike stars. Almost all of these exoplanets lie within a few hundred light-years of the solar system, at distances less than one one-hundredth of the distance to the bulk of the stars in our galaxy.

What does our knowledge of these 5,000 comparatively nearby planets tell us about the possible origin and evolution of life upon them? And what does that imply about the possibilities of widespread life—and even advanced civilizations—in the much vaster reaches of the Milky Way?

On the plus side, our exoplanet discoveries have eliminated the possibility, regarded as viable only a generation ago, that the sites for life's origin in the Milky Way might be few and widely separated. Instead, our galaxy, and presumably others as well, teems with a multitude of diverse cosmic laboratories, some much older than others, in which life might have originated to achieve a currently unknown efflorescence. The immense distances, even within our cosmic neighborhood, eliminate all hope of visiting any of these sites. Instead, we must follow astrophysicists' well-worn path toward greater knowledge, by deriving additional information from better observations.

The likeliest way for life on other worlds to signal its presence (unknowingly, of course) resides in its atmosphere. As we discuss in succeeding chapters of this book, our current understanding and theories about life's origin require the presence of a liquid within which molecules can float and interact. Liquids on the surface of a solid object in turn imply the existence of an atmosphere, whose composition will change from the appearance of life-forms. On Earth, for example, living organisms made our atmosphere oxygen-rich. Civilization has changed the atmosphere in comparatively subtle ways, even as we make heroic escalations.

But how do we distinguish the natural content of a planet's atmosphere from the different composition when life exists there? A clever way uses the time dimension. If we find two or more constituents that should not exist simultaneously, we may deduce that something continuously replenishes them. That something could be alive. Using our own living planet as a guide, the most likely atmospheric components in this detective work would be molecules of oxygen and methane. Without constant resupply, the methane molecules will soon undergo chemical reactions with oxygen (about a hundred thousand times more abundant) and disappear. The fact that our atmosphere maintains a small but detectable amount of methane testifies to living organisms, primarily cows and other ruminants, on Earth's surface, who never cease their release of this gas. (Because methane traps heat much more efficiently than carbon dioxide, even its small but increasing abundance adds to the global warming that threatens our environment.)

Astrophysicists' abilities to study the atmospheres of exoplanets have improved notably with the launch of the new space telescope, whose spectroscopic capabilities can reveal additional details about the composition of the planets' atmospheres. One crucial detection, which unfortunately lies beyond the JWST's capabilities, would find oxygen and methane existing simultaneously. The simultaneous existence of these molecules in a faraway world's atmosphere would leave astrophysicists hard-pressed to avoid the conclusion that life similar to Earth's exists there. To be sure, other forms of life might have metabolisms so different from life on Earth that they would never reveal themselves through this channel. Nevertheless, the oxygen-methane interaction ranks high among the phenomena that exoplanet-oriented astrophysics seeks to investigate. For now, the dream abides, requiring capabilities beyond those of the JWST, and awaiting a new generation of spacecraft to analyze exoplanet atmospheres in detail, or to obtain detailed images of exoplanets that could reveal the existence of life directly. Success with such capabilities would mark the sort of accomplishment about which the bards once sang, elevating mere mortals into heroes for the ages.

PART IV
THE ORIGIN
OF LIFE

I3

LIFE IN THE UNIVERSE

Our survey of origins brings us, as we knew it would, to the most intimate and arguably the greatest mystery of all: the origin of life, and in particular of forms of life with which we may someday communicate. For centuries, humans have wondered how we might find other intelligent beings in the cosmos, with whom we might enjoy at least a modest conversation before we pass into history. The crucial clues for resolving this puzzle may appear in the cosmic blueprint of our own beginnings, which includes Earth's origin within the Sun's family of planets, the origin of the stars that provide energy for life, the origin of structure in the universe, and the origin and evolution of the universe itself.

If we could only read this blueprint in detail, it could direct us from the largest to the smallest astronomical situations, from the unbounded cosmos to individual locations where different types of life flourish and evolve. If we could compare the diverse forms of life that arose under various circumstances, we could perceive the rules of life's beginnings, both in general terms and in particular cosmic situations. Today, we know of only one form of life: life on Earth, all of which shares a common origin and uses DNA molecules as the fundamental means of reproducing itself. This fact deprives us of multiple examples of life, relegating to the future a general survey of life in the cosmos, unachievable until the day we begin to discover forms of life beyond our planet.

Things could be worse. We do know a great deal about life's history on our planet. For the time being, we must build on this knowledge, and on what we understand about the chemistry of Earthlife, to derive basic principles about life throughout the universe. To the extent that we can rely on these principles, they will tell us when and where the universe provides, or has provided, the basic requirements for life. In all our attempts to imagine life elsewhere, we must resist falling into the trap of anthropomorphic thinking, our natural tendency to imagine that extraterrestrial forms of life must be much like our own. This entirely human attitude, which arises from our evolutionary and personal experiences here on Earth, restricts our imagination when we attempt to conceive how different life on other worlds may be. Only biologists familiar with the amazing variety and appearance of different forms of life on Earth can confidently extrapolate what extraterrestrial creatures might look like. Their strangeness almost certainly lies beyond the imaginative powers of ordinary humans.

Someday—perhaps next year, perhaps during the coming century, perhaps long after that—we shall either discover life beyond Earth or acquire sufficient data to conclude, as some scientists now suggest, that life on our planet represents a unique phenomenon within our Milky Way galaxy. For now, our lack of information on this subject allows us to consider an enormously broad range of possibilities: We may find life on several objects in the solar system, which would imply that life probably exists within billions of similar planetary systems in our galaxy. Or we may find that Earth alone within our solar system has life, leaving the question of life around other stars open for the time being. Or we may eventually discover that life exists nowhere around other stars, no matter how far and wide we look. In the search for life in the universe, just as in other spheres of activity, optimism feeds on positive results, while pessimistic views grow stronger from negative outcomes. The most recent information that bears upon the chances for life beyond Earth—the discovery that planets are moving in orbit around many of the Sun's neighboring stars—points toward the optimistic conclusion that life may prove relatively abundant in the Milky Way. Nevertheless, great issues remain to be resolved before this conclusion can gain a firmer footing. If, for example, planets are indeed abundant, but almost none of these planets provide the proper conditions for life, then the pessimistic view of extraterrestrial life seems likely to prove correct.

Scientists who contemplate the possibilities of extraterrestrial life often invoke the Drake equation, after Frank Drake, the American astrophysicist who devised it during the early 1960s. The Drake equation provides a helpful framework rather than a rigorous statement of how the physical universe works. The equation usefully organizes our knowledge and ignorance by separating the number that we dearly seek to estimate—the number of places where intelligent life now exists in our galaxy—into a set of terms, each of which describes a necessary condition for intelligent life. These terms include (1) the number of stars in the Milky Way that survive sufficiently long for intelligent life to evolve on planets around them; (2) the average number of planets around each of these stars; (3) the fraction of these planets with conditions suitable for life; (4) the probability that life actually arises on these suitable planets; and (5) the chance that life on such a planet evolves to produce an intelligent civilization, by which astrophysicists typically mean a form of life capable of communicating with ourselves. When we multiply these five terms, we obtain the number of planets in the Milky Way that possess an intelligent civilization at some point in their history. To make the Drake equation yield the number that we seek—the number of intelligent civilizations that exist at any representative time, such as the present—we must multiply this product by a sixth and final term, the ratio of the average lifetime of an intelligent civilization to the total lifetime of the Milky Way galaxy (about 10 billion years).

Each of the Drake equation's six terms in turn requires astronomical, biological, and sociological knowledge. We now have good estimates of the equation's first two terms, and seem likely to obtain a useful estimate of the third before long. On the other hand, terms four and five—the probability that life arises on a suitable planet, and the probability that this life evolves to produce an intelligent civilization—require that we discover and examine various forms of life throughout the galaxy. For now, anyone can argue almost as well as experts can about the value of these terms. What is the probability, for example, that if a planet does have conditions suitable for life, then life will actually begin on that planet? A scientific approach to this question cries out for the study of several planets suitable for life for a few billion years to see how many do produce life. Any attempt to determine the average lifetime of a civilization in the Milky Way likewise requires several billion years of observation, once we have located a sufficiently large number of civilizations to provide a representative sample.

Isn't this a hopeless task? A full solution of the Drake equation indeed lies immensely far in the future—unless we encounter other civilizations that have already solved it, perhaps using ourselves as a data point. But the equation nevertheless provides useful insights for what it takes to estimate how many civilizations exist in our galaxy now. All six terms in the Drake equation resemble one another mathematically in their effect on the total outcome: each of them exerts a direct, multiplying effect on the equation's answer. If, for instance, you assume that 1 in 3 planets suitable for life actually produces life, but later explorations reveal that the odds are actually 1 in 30, you will have overestimated the number of civilizations by a factor of 10, assuming that your estimates for the other terms prove correct.

Relying on what we now know, the first three terms in the Drake equation imply that billions of potential sites for life exist in the Milky Way. (We restrict ourselves to the Milky Way out of modesty, plus our awareness that civilizations in other galaxies will have a much more difficult time in establishing contact with us, or we with them.) If you like, you can engage in soul-searching arguments with your friends, family, and colleagues about the numbers to assign to the remaining three terms, and choose values that will provide your own estimate for the total number of technologically proficient civilizations in our galaxy. If you believe, for example, that most planets suitable for life do produce life, and that most planets with life do evolve intelligent civilizations, you will conclude that billions of planets in the Milky Way produce an intelligent civilization at some point in their timelines. If, on the other hand, you conclude that only one suitable planet in a thousand does produce life, and only one life-bearing planet in a thousand evolves intelligent life, you will have only thousands, not billions, of planets with an intelligent civilization. Does this enormous range of answers—potentially even wider than the examples given here—imply that the Drake equation presents wild and unbridled speculation rather than science? Not at all. This result simply testifies to the Herculean labor that scientists, along with everyone else, face in attempting to answer an extremely complex question on the basis of highly limited knowledge.

The difficulty that we face in estimating the values of the last three terms in the Drake equation highlights the treacherous step that we take whenever we make a sweeping generalization from a single example—or from none at all. We are hard-pressed, for example, to estimate the average lifetime of a civilization in the Milky Way when we do not even know how long our own will last. Must we abandon all faith in our estimates of these numbers? This would emphasize our ignorance while depriving us of the joy of speculation. If, in the absence of data or dogma, we seek to speculate conservatively, the safest course (though one that might eventually prove to be erroneous) rests on the notion that we are not special. Astrophysicists call this crucial assumption the "Copernican principle" after Nicolaus Copernicus, who, in the mid-1500s, placed the Sun in the middle of our solar system, where it turned out to belong. Until then, despite a third-century BC proposal for a Sun-centered universe by the Greek philosopher Aristarchus, the Earth-centered cosmos had dominated popular opinion throughout most of the past two millennia. Codified by the teachings of Aristotle and Ptolemy, and by the preachings of the Roman Catholic Church, this dogma led most Europeans to accept Earth as the center of all creation. This must have appeared self-evident from a look at the heavens and a natural result from God's plan for the planet. Even today, enormous segments of Earth's human population—quite possibly a majority—continue to draw this conclusion from the fact that Earth seemingly remains immobile while the sky turns around us.

Although we have no guarantee that the Copernican principle can guide us correctly in all scientific investigations, it provides a useful counterweight to our natural tendency to think of ourselves as special. Even more significant is that the principle has an excellent track record so far, leaving us humbled at every turn: Earth does not occupy the center of the solar system, nor does the solar system occupy the center of the Milky Way galaxy, nor the Milky Way galaxy the center of the universe. And in case you believe that the edge is a special place, we are not at the edge of anything, either. A wise contemporary attitude therefore assumes that life on Earth likewise follows the Copernican principle. If so, how can life on Earth, its origins, and its components and structure provide clues about life elsewhere in the universe?

In attempting to answer this question, we must digest an enormous array of biological information. For every cosmic data point, gleaned by long observations of objects at enormous distances from us, we know thousands of biological facts. The diversity of life leaves us all, but especially biologists, awestruck on a daily basis. On this single planet Earth, there coexist (among countless other life-forms), algae, beetles, sponges,

jellyfish, snakes, condors, and giant sequoias. Imagine these
seven living organisms lined up next to each other in order
of size. If you didn't know better, you would be challenged to
believe that they all came from the same universe, much less
the same planet. Try describing a snake to somebody who has
never seen one: "You gotta believe me. I just saw this animal
on planet Earth that (1) stalks its prey with infrared detectors,
(2) swallows whole live animals up to five times bigger than its
head, (3) has no arms or legs or any other appendage, yet (4)
can slide along level ground almost as fast as you can walk!"

In contrast to the amazing variety of life on Earth, the con-
stricted vision and creativity of Hollywood writers who imagine
other forms of life is shameful. Of course, the writers probably
blame a public that favors familiar spooks and invaders over
truly alien ones. But with a few notable exceptions, such as the
life-forms in *The Blob* (1958) and in Stanley Kubrick's *2001:
A Space Odyssey* (1968), Hollywood aliens all look remarkably
humanoid. No matter how ugly (or cute) they may be, nearly
all of them have two eyes, a nose, a mouth, two ears, a head,
a neck, shoulders, arms, hands, fingers, a torso, two legs, two
feet—and they can walk. From an anatomical view, these creatures are practically indistinguishable from humans, yet they
are supposed to live on other planets, the products of independent lines of evolution. A clearer violation of the Copernican
principle can hardly be found. Astrobiology—the study of the
possibilities for extraterrestrial life—ranks among the most
speculative of sciences, but astrobiologists can already assert
with confidence that life elsewhere in the universe, intelligent or otherwise, will surely look at least as exotic as some of
Earth's own life-forms, and quite probably more so. When we
assess the chances of life elsewhere in the universe, we must

attempt to shake from our brains the notions that Hollywood has implanted. Not an easy task, but essential if we hope to reach a scientific rather than an emotional estimate of our chances of finding creatures with whom we may someday have a quiet conversation.

14

THE ORIGIN OF LIFE
ON EARTH

The search for life in the universe begins with a deep question: What is life? Astrobiologists will tell you honestly that this question has no simple or generally accepted answer. Not much use to say that we'll know it when we see it. No matter what characteristic we specify to separate living from nonliving matter on Earth, we can always find an example that blurs or erases this distinction. Some or all living creatures grow, move, or decay, but so too do objects that we would never call alive. Does life reproduce itself? So does fire. Does life evolve to produce new forms? So do certain crystals that grow in watery solutions. We can certainly say that you can tell some forms of life when you see them—who could fail to see life in a salmon or an eagle?—but anyone familiar with life in its diverse forms on Earth will admit that many creatures will remain entirely undetected until the luck of time and the skill of an expert reveal their living nature.

Since life is short, we must press onward with a rough-and-ready, generally appropriate criterion for life. Here it is: life consists of sets of objects that can both reproduce and evolve. We shall not call a group of objects alive simply because they make more of themselves. To qualify as life, they must also evolve into new forms as time passes. This definition therefore eliminates the possibility that any single object can be judged to be alive. Instead, we must examine a range of objects in space and follow them through time. This definition of life may yet prove too restrictive, but for now we shall employ it.

As biologists have examined the different types of life on our planet, they have discovered a general property of Earthlife. The matter within every living Earth creature mainly consists of just four chemical elements: hydrogen, oxygen, carbon, and nitrogen. All the other elements together contribute less than 1 percent of the mass of any living organism. The elements beyond the big four include small amounts of phosphorus, which ranks as the most important, and is essential to most forms of life, together with still smaller amounts of sulfur, sodium, magnesium, chlorine, potassium, calcium, and iron.

But can we conclude that this elemental property of life on Earth must likewise describe other forms of life in the cosmos? Here we can apply the Copernican principle in full vigor. The four elements that form the bulk of life on Earth all appear on the short list of the universe's six most abundant elements. Since the other two elements on that list, helium and neon, almost never combine with anything else, life on Earth consists of the most abundant and chemically active ingredients in the cosmos. Of all the predictions that we can make about life on other worlds, the surest seems to be that their life will be made of elements nearly the same as those used by life on Earth. If life on our planet consisted primarily of four extremely rare elements in the cosmos, such as niobium, bismuth, gallium, and plutonium, we would have an excellent reason to suspect that we represent something special in the universe. Instead, the chemical composition of life on our planet inclines us toward an optimistic view of life's possibilities beyond Earth.

The composition of life on Earth fits the Copernican principle even more than one might initially suspect. If we lived on a planet made primarily of hydrogen, oxygen, carbon, and nitrogen, then the fact that life consists primarily of these four elements would hardly surprise us. But Earth is mainly made of oxygen, iron, silicon, and magnesium, and its outermost layers are mostly oxygen, silicon, aluminum, and iron. Only one of these elements, oxygen, appears on the list of life's most abundant elements. When we look into Earth's oceans, which are almost entirely hydrogen and oxygen, it is surprising that life lists carbon and nitrogen among its most abundant elements, rather than chlorine, sodium, sulfur, calcium, or potassium, which are the most common elements dissolved in seawater. The distribution of the elements in life on Earth resembles the composition of the stars far more than that of Earth itself. As a result, life's elements are more cosmically abundant than Earth's—a good start for those who hope to find life in an array of different situations.

Once we have established that the raw materials for life are abundant throughout the cosmos, we may proceed to ask: How often do these raw materials, along with a site on which these materials can collect, and a convenient source of energy such as a nearby star, lead to the existence of life itself? Someday, when we have made a good survey of possible sites for life among the exoplanets in our Sun's neighborhood, we shall have a statistically accurate answer to this question. In the absence of these data, we must take a roundabout path to an answer and ask, How did life begin on Earth?

The origin of life on Earth remains locked in murky uncertainty. Our ignorance about life's beginnings stems in large part from the fact that whatever events made inanimate matter come alive occurred billions of years ago and left no definitive traces behind. For eras more than 4 billion years in the past, the fossil and geological record of Earth's history does not exist. Yet the interval in solar-system history between 4.6 and 4 billion years ago—the first 600 million years after the Sun and its planets had formed—includes the era when most paleobiologists, specialists in reconstructing life that existed during long-vanished epochs, believe that life first appeared on our planet.

The absence of all geological evidence from epochs more than 4 billion years ago arises from motions of Earth's crust, familiarly called continental drift but scientifically known as plate tectonics. These motions, driven by heat that wells up from Earth's interior, continually force pieces of our planet's crust to slide, collide, and ride by or over one another. Plate-tectonic motions have slowly buried everything that once lay on Earth's surface. As a result, we possess few rocks older than 2 billion years, and none more than 3.8 billion years in age. This fact, together with the reasonable conclusion that the most primitive forms of life had little chance of leaving behind fossil evidence, has left our planet devoid of any reliable record of life during Earth's first 1 or 2 billion years. The oldest definite evidence we have for life on Earth takes us back "only" 2.7 billion years into the past, with indirect indications that life did exist more than 1 billion years before then.

Most paleobiologists believe that life must have appeared on Earth at least 3 billion years ago, and quite possibly 4 billion years ago, within the first 600 million years after Earth formed. Their conclusion relies on a reasonable supposition about primitive organisms. Around 4 billion years ago, significant amounts of oxygen began to appear in Earth's atmosphere. We know this from Earth's geological record independently of any fossil remains: oxygen promotes the slow rusting of iron-rich rocks, producing lovely red tones like those of the rocks in Arizona's Grand Canyon. Rocks from the pre-oxygen era show neither any such colors nor other telltale signs of the element's presence.

The appearance of atmospheric oxygen was the greatest pollution ever to occur on Earth. Atmospheric oxygen does more than combine with iron; it also takes food from the (metaphorical) mouths of primitive organisms by combining with all the simple molecules that could otherwise have provided nutrients for early forms of life. As a result, oxygen's appearance in Earth's atmosphere meant that all forms of life had to adapt or die—and also that if life had not appeared by that time, it could never do so thereafter, because the would-be organisms would have nothing to eat: their potential food would have rusted away. Evolutionary adaptation to this pollution worked well in many cases, as all oxygen-breathing animals can testify. Hiding from the oxygen also did the trick. To this day, every animal's stomach, including our own, harbors billions of organisms that thrive in the anoxic environment that we provide, but would die if exposed to air.

What made Earth's atmosphere relatively rich in oxygen? Much of it came from tiny organisms floating in the seas, which released oxygen as part of their photosynthesis. Some oxygen would have appeared even in the absence of life, as ultraviolet light from the Sun broke apart some of the H_2O molecules at the ocean surfaces, releasing hydrogen and oxy-

gen atoms into the air. Wherever a planet exposes significant
amounts of liquid water to starlight, that planet's atmosphere
should likewise gain oxygen, slowly but surely, over hundreds
of millions or billions of years. There too, atmospheric oxygen would prevent life from originating by combining with all
possible nutrients that could sustain life. Oxygen kills! Not
what we usually say about this eighth element on the periodic
table, but for life throughout the cosmos, this verdict appears
accurate: life must begin early in a planet's history, or else the
appearance of oxygen in its atmosphere will put the kibosh on
life forever.

By ancient coincidence, the epoch missing from the geological record that includes the origin of life also includes the so-
called era of bombardment, which covers those critical first few
hundred million years after Earth had formed. All portions
of Earth's surface must then have endured a continual rain
of objects. During those several hundred thousand millennia,
infalling objects as large as the one that made the Meteor Crater in Arizona must have struck our planet several times in
every century, with much larger objects, each several miles in
diameter, colliding with Earth every few thousand years. Each
one of the large impacts would have caused a local remodel-
ing of the surface, so a hundred thousand impacts would have
produced global changes in our planet's topography.

How did these impacts affect the origin of life? Biologists
tell us that they might have triggered both the appearance
and the extinction of life on Earth, not once but many times.
Much of the infalling material during the era of bombardment
consisted of comets, which are essentially large snowballs full

of tiny rocks and dirt. Their cometary "snow" consists of both frozen water and frozen carbon dioxide, familiarly called dry ice. In addition to their snow, grit, and rocks rich in minerals and metals, the comets that bombarded Earth during its first few hundred million years contained many different types of small molecules, such as methane, ammonia, methyl alcohol, hydrogen cyanide, and formaldehyde. These molecules, along with water, carbon monoxide, and carbon dioxide, provide the raw materials for life. They all consist of hydrogen, carbon, nitrogen, and oxygen, and they all represent the first steps in building complex molecules.

Two simple molecules, hydrogen cyanide (HCN) and hydrogen sulfide (H_2S), each made of hydrogen along with just three types of atoms (carbon, nitrogen, and sulfur), may have played a central role in the origin of life on Earth. During the past dozen years, chemists at Cambridge University have demonstrated that shining ultraviolet light on a bath of these molecules can produce some of the precursor molecules that form far more complex nucleic acids (DNA and RNA), along with some of the material to make amino acids, the basis of protein molecules. Since many comets are rich in hydrogen cyanide, and hydrogen sulfide was apparently abundant on the early Earth, we may yet be able to show that most of life's necessities could have been found in just these two molecular types, with chemical reactions among them induced by ultraviolet from the Sun in the era before our atmosphere blocked most of this radiation.

Cometary bombardment therefore appears to have provided Earth with some of the water for its oceans and with material from which life could begin. Life itself might have arrived in these comets, though their low temperatures, typically hundreds of degrees below zero Fahrenheit, argue against the formation of truly complex molecules. But whether or not life arrived with the comets, the largest objects to strike during the era of bombardment might well have destroyed life that had arisen on Earth. Life might have begun, at least in its most primitive forms, in fits and starts many times over, with each new set of organisms surviving for hundreds of thousands or even millions of years, until a collision with a particularly large object wreaked such havoc on Earth that all life perished, only to appear again, and to be destroyed again, after the passage of a similar amount of time.

We can gain some confidence in the fits-and-starts origin of life from two well-established facts. First, life appeared on our planet sooner rather than later, during the first quarter of Earth's lifetime. If life could and did arise within a billion years, perhaps it could do so in far less time. The origin of life might require no more than a few million, or a few tens of millions, of years. Second, we know that collisions between large objects and Earth have, at intervals of time measured in tens of millions of years, destroyed most of the species alive on our planet. The most famous of these, the Cretaceous-Tertiary extinction 66 million years ago, killed all the non-avian dinosaurs, along with two-thirds of the other species. Even this mass extinction fell short of the most extensive one, the Permian-Triassic mass extinction, that destroyed nearly 90 percent of all species of marine life and 70 percent of all terrestrial vertebrate species, 252 million years ago, leaving fungi as the dominant forms of life on land.

The Cretaceous-Tertiary and Permian-Triassic mass extinctions arose from the collisions of Earth with objects one or two dozen miles across. Geologists have found an enormous

66-million-year-old impact crater, coincident in time with the Cretaceous-Tertiary extinction, that stretches across the northern Yucatán Peninsula and the adjoining seabed. A large crater exists with the same age as the Permian-Triassic extinction, discovered off the northwest coast of Australia, but this mass dying might have arisen from something in addition to a collision, perhaps from sustained volcanic eruptions. Even the single example of the Cretaceous-Tertiary dinosaur extinction reminds us of the immense damage to life that the impact of a comet or asteroid can produce. During the era of bombardment, Earth must have reeled not only from this sort of impact but also from the much more serious effects of collisions with objects 50, 100, or even 250 miles in diameter. Each of these collisions must have cleared the decks of life, either completely or so thoroughly that only a tiny percentage of living organisms managed to survive, and they must have occurred far more often than collisions with 10-mile-wide objects do now. Our present knowledge of astronomy, biology, chemistry, and geology points toward an early Earth ready to produce life, and a cosmic environment ready to eliminate it. And wherever a star and its planets have recently formed, intense bombardment by debris left over from the formation process may even now be eliminating all forms of life on those planets.

More than 4 billion years ago, most of the debris from the solar system's formation either collided with a planet or moved into orbits where collisions could not occur. As a result, our cosmic neighborhood gradually changed from a region of continual bombardment to the overall calm that we enjoy today, broken only at multimillion-year intervals by collisions with objects large enough to threaten life on Earth. You can compare the ancient and continuing threat from impacts whenever you look at the full moon. The giant lava plains that create the face of the "man in the Moon" are the result of tremendous impacts some 4 billion years ago, as the era of bombardment ended, whereas the crater named Tycho, 55 miles across, arose from a smaller, but still highly significant, impact that occurred soon after the dinosaurs disappeared from Earth.

We do not know whether life already existed 4 billion years ago, having survived the early impact storm, or whether life arose on Earth only after relative tranquility began. These two alternatives include the possibility that incoming objects seeded our planet with life, either during the era of bombardment or soon afterward. If life began and died out repeatedly while chaos rained down from the skies, the processes by which life originated seem robust, so that we might reasonably expect them to have occurred again and again on other worlds similar to our own. If, on the other hand, life arose on Earth only once, either as home-grown life or as the result of cosmic seeding, its origin may have occurred here by luck.

In either case, the crucial question of how life actually began on Earth, either once or many times over, has no good answer, though speculation on the subject has acquired a long and intriguing history. Great rewards lie in store for those who can resolve this mystery. From Adam's rib to Dr. Frankenstein's monster, authors have invoked for this purpose a mysterious *élan vital* that imbues otherwise inanimate matter with life.

Scientists seek to probe more deeply, with laboratory experiments and examinations of the fossil record that attempt to establish the height of the barrier between inanimate and animate matter, and to find how nature breached this dike. Early scientific discussions about the origin of life imagined the interaction of simple molecules, concentrated in pools or tide ponds, to create more complex ones. In 1871, a dozen years after the publication of Charles Darwin's seminal book *The Origin of Species*, in which he speculated that "probably all of the organic beings which have ever lived on this Earth have descended from some one primordial form," Darwin wrote to his friend Joseph Hooker:

> It is often said that all the conditions for the first produc-
> tion of a living organism are now present, which could
> ever have been present. But if (and oh! what a big if!) we
> could conceive in some warm little pond, with all sorts of
> ammonia and phosphoric salts, light, heat, electricity, &c.,
> present, that a proteine [*sic*] compound was chemically
> formed ready to undergo still more complex changes, at
> the present day such matter would be instantly absorbed,
> which would not have been the case before living crea-
> tures were found.

In other words, when Earth was ripe for life, the basic compounds necessary for metabolism might have existed in sur-
plus, with nothing in existence to eat them (and, as we have
discussed, no oxygen to combine with them and spoil their
chances to serve as food).

From a scientific perspective, nothing succeeds like experiments that can be compared with reality. In 1953, seeking to
test Darwin's conception of the origin of life in ponds or tide
pools, Stanley Miller, who was then a U.S. graduate student
working at the University of Chicago with the Nobel laureate
Harold Urey, performed a famous experiment that duplicated
the conditions within a highly simplified and hypothetical pool
of water on the early Earth. Miller and Urey partly filled a lab-

oratory flask with water and topped the water with a gaseous mixture of water vapor, hydrogen, ammonia, and methane. They heated the flask from below, vaporizing some of the contents and driving them along a glass tube into another flask, where an electrical discharge simulated the effect of lightning. From there the mixture returned to the original flask, completing a cycle that would be repeated over and over during a few days, rather than a few thousand years. After this entirely modest time interval, Miller and Urey found the water in the lower flask to be rich in "organic gunk," a compound of numerous complex molecules, including different types of sugar, as well as two of the simplest amino acids, alanine and guanine.

Because protein molecules consist of 20 types of amino acids arranged into different structural forms, the Miller-Urey experiment takes us, in a remarkably brief time, a significant part of the way from the simplest molecules to the amino-acid molecules that form the building blocks of living organisms. The Miller-Urey experiment also made some of the modestly complex molecules called nucleotides, which provide the key structural element for DNA, the giant molecule that carries instructions for forming new copies of an organism. Even so, a long path remains before life emerges from experimental laboratories. An enormously significant gap, so far unbridged by human experiment or invention, separates the formation of amino acids—even if our experiments produced all 20 of them, which they do not—and the creation of life. Amino-acid molecules have also been found in some of the oldest and least altered meteorites, believed to have remained unchanged for nearly the entire 4.6-billion-year history of the solar system. This supports the general conclusion that natural processes can make amino acids in many different situations. A balanced view of the experimental results finds nothing totally surprising: the simpler molecules found in living organisms form quickly in many situations, but life does not. The key question still remains: How does a collection of molecules, even one primed for life to appear, ever generate life itself?

Since the early Earth had not weeks but many million years in which to bring forth life, the Miller-Urey experimental results seemed to support the tide-pool model for life's beginnings. Today, however, most scientists who seek to explain life's origin consider the experiment to have been significantly limited by its techniques. Their shift in attitude arose not from doubting the test's results, but rather from recognizing a potential flaw in the hypotheses underlying the experiment. To understand this flaw, we must consider what modern biology has demonstrated about the oldest forms of life.

———

Evolutionary biology now relies on careful study of the similarities and differences among living creatures in their molecules of DNA and RNA, which carry the information that tells an organism how to function and how to reproduce. Careful comparison of these relatively enormous and complex molecules has allowed biologists, among whom the great pioneer has been Carl Woese, to create an evolutionary tree of life that records the "evolutionary distances" among various life-forms, as determined by the degrees to which these life-forms have nonidentical DNA and RNA.

The tree of life consists of three great branches, Archaea, Bacteria, and Eukarya, which replace the biological "kingdoms" formerly believed to be fundamental. The Eukarya include every organism whose individual cells have a well-

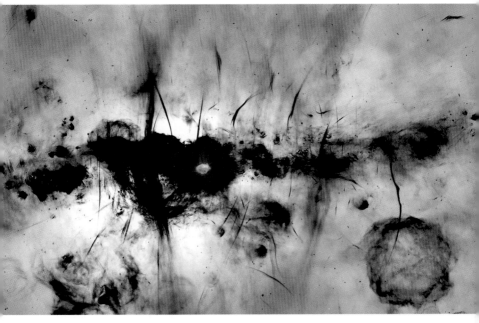

This "Eye of Mordor" image of the central regions of the Milky Way comes from observations made with an array of 64 radio telescopes that stretches over five miles across the semi-desert Karoo regions of South Africa. Large amounts of interstellar dust completely absorb visible light from these regions, about 25,000 light-years from the solar system, but radio waves can penetrate the dust to reveal its details. This image stretches over a few thousand light-years and has color coding to show the strength of radio emission, which becomes most intense in the immediate surroundings of the supermassive black hole at the Milky Way's very center. Other key features are expanding bubbles of gas blown by supernova explosions, with the largest visible at the lower right of the image, and hundreds of long, highly magnetized streams of gas, such as the one that connects to that bubble, which astronomers call "the Snake."

The Crab nebula, the expanding remnants of a supernova explosion observed in AD 1054, lies about 7,000 light-years away, so the actual explosion occurred in about 6,000 BC. In this image, obtained by the Canada-France-Hawaii telescope on Mauna Kea in Hawaii, the reddish filaments consist mainly of hydrogen gas, while the whitish glow arises from electrons that move at nearly the speed of light through intense magnetic fields. Supernova explosions add their material, processed through nuclear fusion, to interstellar clouds of gas and dust that can give birth to new stars (and their planets), which contain more "heavy" elements—such as carbon, nitrogen, oxygen, and iron—than older stars do.

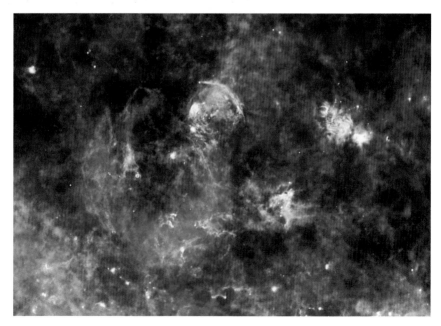

The supernova remnant W44, about 10,000 light-years from the solar system, has ejected material at high speed into the surrounding gas and dust. This image, secured in long-wavelength infrared, shows the expanding shell of ejected material, about 100 light-years in diameter, to the left, while the top of the image reveals this shell impacting a large region of hot gas. All around the supernova remnant lie star-forming regions within which massive hot stars are being born or have just been born, astronomically speaking.

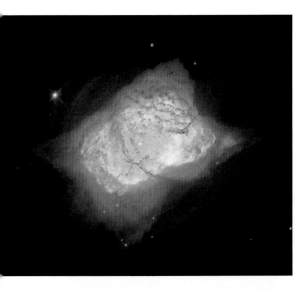

This expanding gaseous region, the expelled remnants of an aging star that has become a white dwarf, spans about one light-year and lies about 3,000 light-years away. The Hubble Space Telescope has made repeated observations of this mass of gas and dust that allow astronomers to measure its expansion.

The Dumbbell nebula, approximately 1,200 light-years from the solar system, likewise consists of gas and dust ejected by an aging star, visible at the nebula's center, whose ultraviolet radiation heats the surrounding gas and makes it glow in visible light.

In northern Chile, the Atacama Large Millimeter Array (ALMA) of 66 dishes profits from the 16,500-foot altitude and desert conditions to observe the cosmos in infrared light that does not penetrate to lower elevations. This image of the young sun-like star TW Hydrae, about 175 light-years away, reveals a surrounding disk of material, banded into regions of greater and lesser concentrations of gas and dust that we see nearly face-on. This disk of material seems likely to form a planet—indeed, some planets may already be orbiting in the dark gaps between the brighter portions of the disk.

This graph of known exoplanets plots their sizes along the vertical axis and their distances from their stars along the horizontal axis. Different discovery techniques have revealed exoplanets with notably different values of these parameters, denoted by the color-coding defined at the lower left of the diagram.

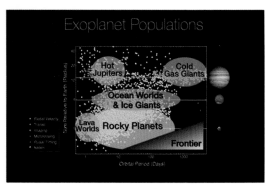

Three gravitational-wave detectors, almost identical in their design and capability, now occupy sites in Italy, Louisiana, and Washington State, awaiting slight ripples from violent cosmic events. These ripples produce tiny changes in the lengths of two perpendicular 2.5-mile-long arms in each location, which can be detected by using mirrors that reflect laser beams hundreds of times back and forth.

The technicians and engineers preparing the James Webb Space Telescope for shipment from California to its launch site in French Guiana give a sense of the telescope's size, as well as the complex folding of the telescope's 18 mirror segments and its sunshield that allowed the telescope to fit into the spacecraft for launch.

On December 25, 2021, ESO's Ariane 5 rocket carried the JWST into space. A few months later, after reaching the Earth-Sun L2 orbital position, about a million miles from Earth in the direction opposite to the Sun, the telescope completed its preparations and began to study the cosmos with a precision never before possible.

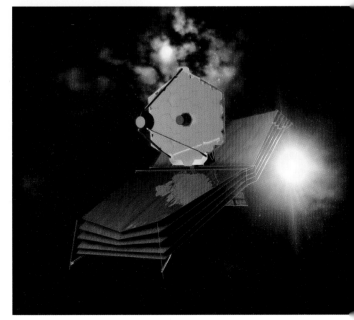

In May 2019, the experienced photographer Matthew Vandeputte, after setting up his camera to take photographs while he slept, awoke to find that he had captured a "bolide," a particularly bright meteor that explodes as its trajectory ends high in the atmosphere. The trail left behind by this interplanetary interloper lingered for half an hour before dissipating.

In August 2014, the European Space Agency's Rosetta spacecraft and Philae lander achieved the first landing upon a comet. Made from a mixture of rocky material with frozen water (ice) and frozen carbon dioxide ("dry ice"), this comet generally resembles others, but comes unusually close to Earth, though never closer than about 27 million miles.

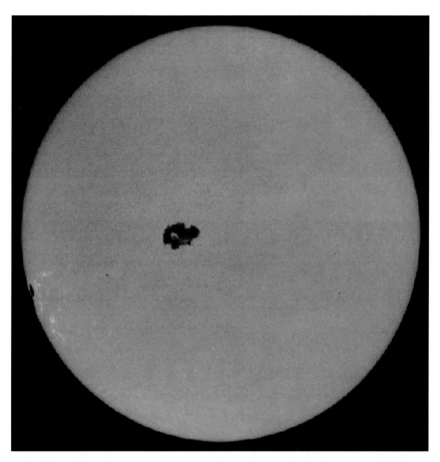

This large sunspot, ten times the size of Earth, appears dark because its gases, with a temperature of about 8,000 degrees Fahrenheit, are cooler than the 10,000 degrees Fahrenheit temperature of the rest of the solar photosphere (the region of the sun's gaseous layers from which its light emerges).

On July 31, 1971, two of the Apollo 15 astronauts became the first humans to explore the Moon with an electric vehicle designed for non-aerodynamic, low-gravity operation.

In 1990, the Magellan orbiter secured this radar image of the surface of Venus, a close twin of Earth in size and mass but with an atmosphere a hundred times thicker than Earth's. Made primarily of carbon dioxide, but with additional particles that make it opaque, Venus's atmosphere traps heat and maintains nearly constant surface temperatures close to 900 degrees Fahrenheit.

On Mars, barely more than half of Earth's diameter, the tallest volcanoes and the longest and deepest canyon in the solar system are seen through a thin atmosphere of carbon dioxide, whose frozen state ("dry ice") forms the polar caps. Although exposed liquid water quickly evaporates, evidence suggests that large pools of water may lurk beneath the planet's surface.

Jezero Crater on Mars, a small portion of which appears in this image, offers a fine place to search for evidence of ancient microbial life, especially at the delta of the river that flowed into the crater about 3.5 billion years ago. *Perseverance* has been exploring this area of Mars since it landed on the Red Planet in February 2021.

In April 2021, the *Ingenuity* helicopter became the first aircraft to fly through Mars's thin atmosphere. Actually, it was the first aircraft to fly anywhere in the solar system other than Earth. This photograph shows the *Perseverance* rover at the extreme left.

The *Curiosity* rover, which has been active on Mars since 2012, obtained this 360-degree panorama (shown here compressed in its horizontal dimension), as it climbed up the side of Mount Sharp, the central peak in Gale Crater. The vista includes hills with geologically layered gray sand and red rocky ground, with notable dust in the thin atmosphere.

In 2022, the *Perseverance* rover secured this mosaic image of the hill named "Santa Cruz," about 165 feet tall and 1.6 miles away, which is located near the edge of Jezero Crater's delta. The boulders seen in the foreground are about 20 inches across.

In 2015, NASA's *Dawn* spacecraft approached within 8,000 miles of Ceres, the largest solar-system asteroid with a diameter of 590 miles, although elevated to the status of dwarf planet in 2006. Its heavily cratered surface resembles our Moon, which is almost four times wider than Ceres.

Since 2016, NASA's *Juno* spacecraft has orbited Jupiter, securing highly detailed photographs of the giant planet's changing atmospheric layers, measuring its magnetic field, and obtaining other relevant measurements. In this image, the colors have been boosted for clarity.

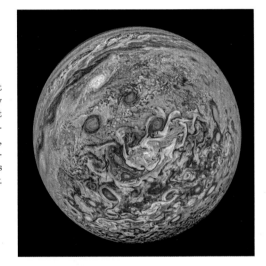

Jupiter's Europa, about the size of Earth's Moon, has an icy surface, with long, thin fractures that stretch for hundreds of miles. Future spacecraft could drill through the ice to test the moonwide ocean of liquid water, kept warm by tidal flexing, which might harbor extraterrestrial organisms.

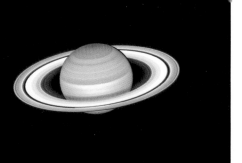

This Hubble Space Telescope photograph of Saturn, taken in 2020, shows the magnificent rings that span a distance more than 20 times the Earth's diameter. Two small moons appear in this image: Mimas to the right of Saturn, and Enceladus directly below the planet.

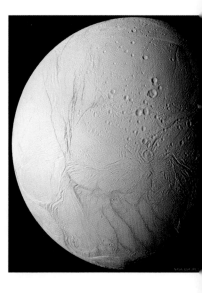

Saturn's moon Enceladus, just 313 miles in diameter, has an ice-covered surface that displays long features, called "tiger stripes," that spew ice into space to form Saturn's outermost ring.

Saturn's giant moon Titan, in a photo-finish with Jupiter's Ganymede as the largest in the solar system, has a thick atmosphere, made mainly of nitrogen (like Earth's), but infused with smog that forever hides the moon's surface from view.

Titan's surface, much too cold for liquid water to exist, has numerous lakes of liquid methane, potential sites for life much different from what we have on Earth.

On July 14, 2015, the *New Horizons* spacecraft obtained the first detailed images of Pluto's surface. On this dwarf planet, temperatures hundreds of degrees below zero allow ice mountains to stand above nearly flat basins.

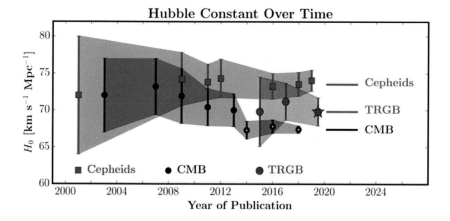

Hubble Constant Over Time

Cepheids

TRGB

CMB

- ■ Cepheids
- ● CMB
- ● TRGB

Year of Publication

This graph shows three different methods to determine the Hubble constant over time, as better observations and data analysis have improved the results. The widths of the colored zones show the likely uncertainties associated with the derived values. The method denoted in blue and called "Cepheids" uses variable stars to benchmark distances to faraway Type Ia supernovae, while the method in red, with the abbreviated name TRGB, instead uses stars in star clusters to set those benchmarks. A completely different method, denoted in charcoal, relies on the cosmic background radiation (CBR), as shown in Plate 3.

The Tadpole galaxy, more formally known as Arp 188, lies 420 million light-years from the Milky Way. This Hubble Space Telescope photograph shows the galaxy's lengthy "tail," approximately 280,000 light-years long, which arose from the gravitational force exerted by a passing galaxy, faintly visible through the spiral arms at the top of the Tadpole's image. The encounter drew stars, gas, and dust into the tail, within which star clusters will gradually become satellites of the large spiral.

Gravitational Lens G2237+0305

The Hubble Space Telescope obtained this image of the "gravitational lens" designated as G2237+0304, familiarly called the Einstein Cross. A chance line-up has caused the light from a quasar 8 billion light-years from the Milky Way to be diverted along four different pathways by the gravitational force from an intervening, relatively nearby galaxy, 400 million light-years away, which appears as a diffuse object at the center of the cross.

The galaxy cluster Abell 370, about 4 billion light-years from our galaxy, has several hundred galaxies of different types, whose mutual gravitational interactions have allowed the cluster to have a roughly spherical shape. The bluish arcs arise from far more distant galaxies, too faint to appear in the image, but whose light, in some cases, has been diverted and magnified by the gravitational force from one of the galaxies in the cluster.

defined center or nucleus that contains the genetic material governing the cells' reproduction. This characteristic makes Eukarya more complex than the other two types. Every form of life familiar to the nonexpert belongs to this branch. We may reasonably conclude that Eukarya arose later than Archaea or Bacteria. And because Bacteria lie further from the origin of the tree of life than the Archaea do—for the simple reason that their DNA and RNA have changed more—the Archaea, as their name implies, almost certainly represent the oldest forms of life. Now comes a shocker: unlike the Bacteria and Eukarya, the Archaea consist mainly of "extremophiles," organisms that love to live, and live to love, in what we now call extreme conditions, defined by temperatures near or above the boiling point of water, high acidity, or other situations that would kill other forms of life. (Of course, if the extremophiles had their own biologists, they would classify themselves as normal and any life that thrives at room temperature as an extremophile.) Modern research into the tree of life tends to suggest that life began with the extremophiles, and only later evolved into forms of life that benefit from what we call normal conditions.

In that case, Darwin's "warm little pond," as well as the tide pools duplicated in the Miller-Urey experiment, would evaporate into the mist of rejected hypotheses. Gone would be the relatively mild cycles of drying and wetting. Instead, those who seek to find the places where life may have begun would have to look to locales where extremely hot water, possibly rich in acids, surges from Earth.

The past few decades have allowed oceanographers to discover just such places, along with the strange forms of life they support. In 1977, two oceanographers piloting a deep-sea submersible vehicle discovered the first deep-sea vents, a mile and a half beneath the calm surface of the Pacific Ocean near the Galápagos Islands. At these vents, Earth's crust behaves locally like a household cooker, generating high pressure inside a heavy-duty pot with a lockable lid and heating water beyond its ordinary boiling temperature without letting it reach an actual boil. As the lid partially lifts, the pressurized, super-heated water spews out from below Earth's crust into the cold ocean basins.

The superheated seawater that emerges from these vents carries dissolved minerals that quickly collect and solidify to surround the vents with giant, porous rock chimneys, hottest in their cores and coolest at the edges that make direct contact with seawater. Across this temperature gradient live count-less life-forms that have never seen the Sun and care nothing for solar heating, though they do require the oxygen dissolved in seawater, which in turn comes from the existence of solar-driven life near the surface. These hardy bugs live on geo-thermal energy, which combines heat left over from Earth's formation with heat continuously produced by the radioactive decay of unstable isotopes such as aluminum-26, which lasts for millions of years, and potassium-40, which lasts for bil-lions. Despite their enormously different systems for obtaining and using energy, these organisms depend on DNA to govern their biological processes and reproduction, just as all other forms of Earthlife do.

Near these vents, far below the depths to which any sun-light can penetrate, the oceanographers found tube worms as long as a person, thriving amid large colonies of bacteria and other small creatures. Instead of drawing its energy from sun-light, as plants do with photosynthesis, life near deep-sea vents relies on "chemosynthesis," the production of energy by chemical reactions, which in turn depend on geothermal heating.

How does this chemosynthesis occur? The hot water gushing from the deep-sea vents emerges laden with hydrogen-sulfur and hydrogen-iron compounds. Bacteria near the vents combine these molecules with the hydrogen and oxygen atoms in water molecules, and with the carbon and oxygen atoms of the carbon dioxide molecules dissolved in seawater. These reactions form larger molecules—carbohydrates—from carbon, oxygen, and hydrogen atoms. Thus the bacteria near deep-sea vents mimic the activities of their cousins far above, which likewise make carbohydrates from carbon, oxygen, and hydrogen. One set of microorganisms draws the energy to make carbohydrates from sunlight, and the other from chemical reactions at the ocean floors. Close by the deep-sea vents, other organisms consume the carbohydrate-making bacteria, profiting from their energy in the same way that animals eat plants, or eat plant-eating animals.

In the chemical reactions near deep-sea vents, however, more goes on than the production of carbohydrate molecules. The iron and sulfur atoms, which are not included in carbohydrate molecules, combine to make compounds of their own, most notably crystals of iron pyrite, familiarly called "fool's gold," known to the ancient Greeks as "fire stone" because a good blow from another rock will strike sparks from it. Iron pyrite, the most abundant of all the sulfur-bearing minerals found on Earth, might have played a crucial role in the origin of life by encouraging the formation of carbohydrate-like molecules. This hypothesis sprang from the mind of a German patent attorney and amateur biologist, Günter Wächtershäuser, whose profession hardly excludes him from biological speculation, any more than Einstein's work as a patent attorney barred him from insights into physics. (To be sure, Einstein had an advanced degree in physics, while Wächtershäuser's biology and chemistry are mainly self-taught.)

In 1994, Wächtershäuser proposed that the surfaces of iron pyrite crystals, formed naturally by combining iron and sulfur that surged from deep-sea vents early in Earth's history, would have offered natural sites where carbon-rich molecules could accumulate, acquiring new carbon atoms from the material ejected by the nearby vents. Like those who hypothesize that life began in ponds or tide pools, Wächtershäuser has no clear way to pass from the building blocks to living creatures. Nevertheless, with his emphasis on the high-temperature origin of life, he may prove to be on the right track, as he firmly believes. Referring to the highly ordered structure of iron pyrite crystals, on whose surfaces the first complex molecules for life might have formed, Wächtershäuser has confronted his critics at scientific conferences with the striking statement that "some say that the origin of life brings order out of chaos—but I say, 'order out of order out of order!'" Delivered with German brio, this claim acquires a certain resonance, though only time can tell how accurate it may be.

So which basic model for life's origin is more likely to prove correct—tide pools at the ocean's edge, or superheated vents on the ocean floors? For now, the betting is about even. Experts on the origin of life have challenged the assertion that life's oldest forms lived at high temperatures, because current methods for placing organisms at different points along the branches of the tree of life remain the subject of debate. In addition, computer programs that trace out how many compounds of different types existed in ancient RNA molecules, the close cousins of DNA that apparently preceded DNA in life's history, suggest that the compounds favored by high temperatures appeared only after life had undergone some relatively low-temperature history.

Thus the outcome of our finest research, as so often occurs in science, proves unsettling to those who seek certainty. Although we can state approximately when life began on Earth, we don't know where or how this marvelous event occurred. Paleobiologists have recently given the elusive ancestor of all Earthlife the name LUCA, the "last universal common ancestor." (See how firmly these scientists' minds have remained fixed to our planet: they should call life's progenitor LECA, for the last earthly common ancestor.) For now, naming this ancestor—a set of primitive organisms that all shared the same genes—mainly underscores the distance that we still must travel before we can pierce the veil that separates life's origin from our understanding.

More than a natural curiosity as to our own beginnings hinges on the resolution of this issue. Different origins for life imply different possibilities for its origin, evolution, and survival both here and elsewhere in the cosmos. For example, Earth's ocean floors may provide the most stable ecosystem on our planet. If a jumbo asteroid slammed into Earth and rendered all surface life extinct, the oceanic extremophiles would almost certainly continue undaunted in their happy ways. They might even evolve to repopulate Earth's surface after each extinction episode. And if the Sun were mysteriously plucked from the center of the solar system and Earth drifted through space, this event would hardly merit attention in the extremophile press, as life near deep-sea vents might continue relatively undisturbed. But in 5 billion years, the Sun will become a red giant as it expands to fill the inner solar system. Meanwhile, Earth's oceans will boil away and Earth itself will partially vaporize. Now that would be news for any form of Earthlife.

The ubiquity of extremophiles on Earth leads us to a profound question: Could life exist deep within many of the rogue planets or planetesimals that were ejected from the solar system during its formation? Their "geo"thermal reservoirs could last for billions of years. What about the countless planets that were forcibly ejected by every other solar system that ever formed? Could interstellar space be teeming with life—formed and evolved deep within these starless planets? Before astrophysicists recognized the importance of extremophiles, they envisioned a "habitable zone" surrounding each star, within which water or another substance could maintain itself as a liquid, allowing molecules to float, interact, and produce more complex molecules. Today, we must modify this concept, so that, far from being a tidy region around a star that receives just the right amount of sunlight, a habitable zone can be anywhere and everywhere, maintained not by starlight heating but by localized heat sources, often generated by radioactive rocks. So the Three Bears' cottage was, perhaps, not a special place among fairy tales. Anybody's residence, even one of the Three Little Pigs', might contain a bowl of food at a temperature that is just right.

What a hopeful, even prescient, fairy tale this may prove to be. Life, far from being rare and precious, may be almost as common as planets themselves. All that remains is for us to go find it.

15

SEARCHING FOR LIFE IN THE SOLAR SYSTEM

The possibility of life beyond Earth has created new job titles, applicable to only a few individuals but potentially capable of sudden growth. Astrobiologists grapple with the issues presented by life beyond Earth, whatever forms that life may take. For now, astrobiologists can only speculate about extraterrestrial life or simulate extraterrestrial conditions, to which they either expose terrestrial life-forms, testing how they may survive harsh and unfamiliar situations, or subject mixtures of inanimate molecules, creating a variant on the classic Miller-Urey experiment or a gloss on Wächtershäuser's research. This combination of speculation and experiment has led them to several generally accepted conclusions, which—to the extent that they describe the real universe—have highly significant implications. Astrobiologists now think that the existence of life throughout the universe requires

1. a source of energy;
2. a type of atom that allows complex structures to exist;
3. a liquid solvent in which molecules can float and interact; and
4. sufficient time for life to arise and to evolve.

On this short list, requirements (1) and (4) present only low barriers to the origin of life. Every star in the cosmos provides a source of energy, and all but the most massive 1 percent of these stars last for hundreds of millions or billions of years. Our Sun, for example, has furnished Earth with a steady supply of heat and light during the past 5 billion years, and will continue to do so for another 5 billion. Furthermore, we now see that life can exist entirely without sunlight, relying on geothermal heating and chemical reactions for its energy. Geothermal energy arises in part from the radioactivity of isotopes of elements such as potassium, thorium, and uranium, whose decay occurs over time scales measured in billions of years—a time scale comparable to the lifetime of all Sunlike stars.

––––––––

On Earth, life satisfies point (2), the requirement of a structure-building atom, with the element carbon. Carbon atoms can each bind to one, two, three, or four other atoms, which makes them the crucial element in the structure of all the life we know. In contrast, hydrogen atoms can each bind to only one other atom, and oxygen to only one or two. Because carbon atoms can bind with as many as four other atoms, they form the "backbone" for most of the molecules within living organisms, such as proteins and sugars.

Carbon's ability to create complex molecules has made it one of the four most abundant elements, together with hydrogen, oxygen, and nitrogen, in all forms of life on Earth. We have seen that although the four most abundant elements in Earth's crust have only one match with these four, the universe's six most abundant elements include all four of those in Earthlife, along with the inert gases helium and neon. This fact could support the hypothesis that life on Earth began in the stars, or in objects whose composition resembles those of the stars. In any case, the fact that carbon forms a relatively small fraction of Earth's surface but a large part of any living creature testifies to carbon's pivotal role in giving structure to life.

Is carbon essential to life throughout the cosmos? What about the element silicon, which often appears in science-fiction novels as the basic structural atom for exotic forms of life? Like carbon, silicon atoms bond with as many as four other atoms, but the nature of these bonds leaves silicon far less likely than carbon to provide the structural basis for complex molecules. Carbon bonds to other atoms rather weakly, so that carbon-oxygen, carbon-hydrogen, and carbon-carbon bonds, for example, break with relative ease. This allows carbon-based molecules to form new types as they collide and interact, an essential part of any life-form's metabolic activity. In contrast, silicon forms strong bonds with many other types of atoms, and in particular with oxygen. Earth's crust consists largely of silicate rocks made primarily of silicon and oxygen atoms, bound together with sufficient strength to last for millions of years, and therefore unavailable to participate in forming new types of molecules at temperatures that sustain liquid water.

The difference between the way that silicon and carbon atoms bond to other atoms argues strongly that we may expect to find most, if not all, extraterrestrial life-forms built, as we are, with carbon, not silicon, backbones for their molecules. Other than carbon and silicon, only relatively exotic types of atoms, with cosmic abundances much lower than those of carbon or silicon, can bond with as many as four other atoms.

Purely on numerical grounds, the possibility that life uses atoms such as germanium in the same way that Earthlife uses carbon seems highly remote.

––––––––

Requirement number (3) specifies that all forms of life need a liquid solvent in which molecules can float and interact. The word "solvent" emphasizes that a liquid allows this float-and-interact situation, in what chemists call a "solution." Liquids allow relatively high concentrations of molecules but do not place tight restrictions on their motions. In contrast, solids lock atoms and molecules in place. Although they can actually collide and interact, they do so far more slowly than in liquids. In gases, molecules will move even more freely than in liquids, and can collide with even less hindrance, but their collisions and interactions occur far less often than they do in liquids, because the density within a liquid typically exceeds that within a gas by a factor of 1,000 or more. "Had we but world enough and time," as Andrew Marvell wrote, we might find life originating in gases rather than liquids. In the real cosmos, only 14 billion years old, astrobiologists do not expect to find life that began in gas. Instead, they expect all extraterrestrial life, like all life on Earth, to consist of sacs of liquid, within which complex chemical processes occur as different types of molecules collide and form new types.

Must that liquid be water? We live on a watery planet whose oceans cover nearly three-quarters of the surface. This makes us unique in our solar system, and possibly a highly unusual planet anywhere in our Milky Way galaxy. Water, which consists of molecules made from two of the most abundant elements in the cosmos, appears at least in modest amounts in comets, in meteoroids, and in most of the Sun's planets and their moons. On the other hand, liquid water in the solar system exists only on Earth and beneath the icy surface of Jupiter's large moon Europa, whose worldwide covered ocean remains only a likelihood, not a verified reality. Could other compounds offer better chances for liquid seas or ponds, within which molecules could have found their way to life? The three most abundant alternative compounds that can remain liquid within a significant range of temperatures are ammonia, ethane, and methyl alcohol. Ammonia molecules each consist of three hydrogen atoms and one nitrogen atom, ethane of two hydrogen atoms and six carbon atoms, and methyl alcohol of four hydrogen atoms, one carbon atom, and one oxygen atom. When we consider the possibilities for extraterrestrial life, we may reasonably consider creatures that use ammonia, ethane, or methyl alcohol in the way that Earthlife employs water—as the fundamental liquid within which life presumably originated, and which supplies the medium within which molecules can amble their way to glory. The Sun's four giant planets possess enormous amounts of ammonia, along with smaller amounts of methyl alcohol and ethane, and Saturn's large moon Titan has lakes of liquid ethane on its frigid surface.

The choice of a particular type of molecule as life's basic liquid immediately implies another requirement for life: the substance must remain liquid. We would not expect life to originate in the Antarctic ice cap, or in clouds rich in water vapor, because we need liquids to allow abundant molecular interactions. Under atmospheric pressures like those at Earth's surface, water remains liquid between 0 and 100 degrees Celsius (32°F–212°F). All three of the alternative types of solvents remain liquid within temperature ranges that extend far below water's. Ammonia, for example, freezes at −78 degrees Celsius and vaporizes at −33 degrees. This prevents ammonia from providing a liquid solvent for life on Earth, but on a world with a temperature 75 degrees colder than ours, where water could never serve as a solvent for life, ammonia might well be the charm.

———————

Water's most significant distinguishing feature does not consist of its well-earned badge of "universal solvent," about which we learned in chemistry class, nor of the wide temperature range over which water remains liquid. Its most remarkable attribute resides in the fact that while most things—water included—shrink and become denser as they cool, water that cools below 4 degrees Celsius expands, becoming progressively less dense as the temperature falls toward zero. And then, when water freezes at 0 degrees Celsius, it turns into an even less dense substance than liquid water. Ice floats, which is very good news for fish. During the winter, as the temperature of the outside air drops below freezing, 4-degree water sinks to the bottom and stays there, because it is denser than the colder water above, while a floating layer of ice builds slowly on the surface, insulating the warmer water below.

Without this density inversion below 4 degrees, ponds and lakes would freeze from the bottom up, not from the top down. Whenever the outside air temperature fell below freezing, a pond's upper surface would cool and sink to the bottom as warmer water rose from below. This forced convection would rapidly drop the water's temperature to 0 degrees Celsius as the surface began to freeze. Then denser, solid ice would sink to the bottom. If the entire body of water did not freeze from the bottom upward in a single season, the accumulation of ice at the bottom would allow full freezing to occur over the course of many years. In such a world, the sport of ice fishing would yield even fewer results than it does now, because all the fish would be dead—fresh-frozen. Ice anglers would find themselves on a layer of ice that was either submerged below all remaining liquid water or atop a completely frozen body of water. No longer would you need icebreakers to traverse the frozen Arctic—either the entire Arctic Ocean would be frozen solid, or the frozen parts would all have sunk to the bottom and you could just sail your ship without incident. You could slip and slide on lakes and ponds without fear of falling through. In this altered world, ice cubes and icebergs would sink, so that in April 1912, the *Titanic* would have steamed safely into the port of New York City, unsinkable (and unsunk) as advertised.

On the other hand, our mid-latitude prejudice may be showing here. Most of Earth's oceans are in no danger of freezing, whether from the top down or the bottom up. If ice sank, the Arctic Ocean might become solid, and the same might happen to the Great Lakes and the Baltic Sea. This effect could have inhibited travel and communications within Europe and the United States, but life on Earth would have persisted and flourished just as well.

Let us, for the time being, adopt the hypothesis that water has such significant advantages over its chief rivals, ammonia and methyl alcohol, that most, if not all, forms of extraterrestrial life must rely on the same solvent that Earthlife does. (We can widen our horizons when we consider the possibilities of life on Titan or even—amazingly—on Pluto.) Armed with this supposition, along with the general abundance of the raw

materials for life, the prevalence of carbon atoms, and the long
stretches of time available for life to appear and to evolve, let
us take a tour of our neighbors, recasting the age-old question,
Where's the life? into the more modern one, Where's the water?

———

If you were to judge matters by the appearance of some dry
and unfriendly-looking places in our solar system, you might
conclude that water, while plentiful on Earth, ranks as a rare
commodity elsewhere in our galaxy. But of all the molecules
that can be formed with three atoms, water is by far the most
abundant, largely because water's two constituents, hydrogen
and oxygen, occupy positions one and three on the abundance
list. This suggests that rather than asking why some objects
have water, we should ask why they don't all possess large
amounts of this simple molecule.

How did Earth acquire its oceans of water? The Moon's near-
pristine record of craters tells us that impacting objects have
struck the Moon throughout its history. We may reasonably
expect that Earth has likewise undergone many collisions.
Indeed, Earth's larger size and stronger gravity imply that
we should have been struck many more times, and by larger
objects, than the Moon, one of which, as described in Chapter
11, gave birth to the Moon itself. So it has been, from its birth
all the way to the present. After all, Earth didn't hatch from
an interstellar void, springing into existence as a preformed
spherical blob. Instead, our planet grew within the condensing
gas cloud that formed the Sun and its other planets. In this
process, Earth assembled itself by accreting enormous numbers of small solid particles, and eventually through incessant
impacts from mineral-rich asteroids and water-rich comets.

How incessant? The early impact rate of comets may have been sufficiently large to have brought us the water in all our oceans. Uncertainties (and controversies) continue to surround this hypothesis. The water that we observed in comet Halley has far greater amounts than Earth does of deuterium, an isotope of hydrogen that packs an extra neutron into its nucleus. If Earth's oceans arrived in comets, then those that hit Earth soon after the solar system formed must have had a chemical composition notably different from today's comets, or at least different from the class of comet from which Halley is drawn.

In any case, when we add their contribution to the water vapor spewed into the atmosphere by volcanic eruptions, we have no shortage of pathways by which Earth could have acquired its supply of surface water.

————————

If you seek a waterless, airless place to visit, you need look no farther than Earth's Moon. The Moon's near-zero atmospheric pressure, combined with its two-week-long days when the temperature rises to 200 degrees Fahrenheit, causes any water to evaporate swiftly. During the two-week lunar night, the temperature can drop to 250 degrees below zero, sufficient to freeze practically anything. The Apollo astronauts who visited the Moon therefore brought all the water and air (and the air-conditioning) that they needed for their round-trip journey.

It would be odd, however, if Earth had acquired a great deal of water while the nearby Moon got almost none. One possibility, certainly true at least in part, is that water evaporated from the Moon's surface much more readily than from Earth's because of the Moon's lesser gravity. Another possibility suggests that lunar missions may eventually not need to import water or the assortment of products derived from it. Observations by the *Clementine* lunar orbiter, which carried an instrument to detect the neutrons produced when fast-moving interstellar particles collide with hydrogen atoms, support a long-held contention that deep-frozen ice deposits may lurk beneath craters near the Moon's north and south poles. If the Moon receives an average number of impacts per year from interplanetary flotsam, then the mixture of these impactors should, from time to time, include sizable water-rich comets, like those that strike Earth. How big could these comets be? The solar system contains plenty of comets that could melt into a puddle the size of Lake Erie.

While we can't expect a freshly laid lake to survive many sunbaked lunar days at temperatures of 200 degrees, any comet that happened to crash in the bottom of a deep crater near one of the Moon's poles (or happened to make a deep polar crater itself) would remain shrouded in darkness, because deep craters near its poles are the only places on the Moon where the "Sun don't shine." If you thought that the Moon has a perpetual dark side, you have been badly misled by many sources, probably including Pink Floyd's 1973 album *Dark Side of the Moon*. As light-starved Arctic and Antarctic dwellers know, the Sun in those regions never rises high in the sky at any time of day or any season of the year. Now imagine living at the bottom of a crater whose rim rises higher than the highest altitude that the Sun ever reaches. With no air to scatter sunlight into the shadows, you would live in eternal darkness.

But even in cold darkness, ice slowly evaporates. Just look at the cubes in your freezer's ice tray upon your return from a long vacation: their sizes will be distinctly smaller than when you departed. However, if ice has been well mixed with solid particles (as occurs in a comet), it can survive for thousands and millions of years at the bottom of the Moon's deep polar craters. Any outpost that we might establish on the Moon would benefit greatly from being located near this lake. Apart from the obvious advantages of having ice to melt, to filter, and then to drink, we could also profit by dissociating the water's hydrogen from its oxygen atoms. We could use the hydrogen, plus some of the oxygen, as active ingredients for rocket fuel, while keeping the rest of the oxygen for breathing. And in our spare time between space missions, we might choose to go skating.

———

Although Venus has nearly the same size and mass as Earth, several attributes distinguish our sister planet from all the other planets in the solar system, including its highly reflective, thick, dense, carbon dioxide atmosphere, which exerts 92 times the surface pressure of Earth's atmosphere. Except for bottom-dwelling marine creatures that live at similar pressures, all forms of Earthlife would be crushed to death on Venus. Venus's most peculiar feature, however, resides in the relatively young craters uniformly scattered over its surface. This innocuous-sounding description implies that a recent planetwide catastrophe reset the cratering clock—and thus our ability to date a planet's surface by its buildup of craters—by wiping out the evidence of all previous impacts. A major erosive weather phenomenon such as a planetwide flood might also have done this. But so could planetwide geologic (should we say Venusologic?) activity, such as lava flows, which could have turned Venus's entire surface into the American automotive dream—a totally paved planet. Whatever events reset the cratering clock must have ceased abruptly. But important questions remain, in particular about Venus's water. If a planetwide flood did occur on Venus, where has all the water gone? Did it sink below the surface? Did it evaporate into the atmosphere? Or did the flood consist of a common substance other than water? Even if no flood occurred, Venus presumably acquired about as much water as its sister planet Earth. What has happened to it?

The answer seems to be that Venus lost its water by growing too hot, a result attributable to Venus's atmosphere. Although carbon dioxide molecules let visible light pass by them, they trap infrared radiation with great efficiency. Sunlight can therefore penetrate Venus's atmosphere, even though atmospheric reflection reduces the amount of sunlight that reaches the surface. This sunlight heats the planet's surface, which radiates infrared, and which cannot escape. Instead, the carbon dioxide molecules trap it, as the infrared radiation heats the lower atmosphere and the surface below. Scientists call this trapping of infrared radiation the "greenhouse effect" by loose analogy to greenhouses' glass windows, which admit visible light but block some of the infrared. Like Venus and its atmosphere, Earth produces a greenhouse effect, essential for many forms of life, that raises our planet's temperature by about 25 degrees Fahrenheit over what we would find in the absence of an atmosphere. Most of our greenhouse effect arises from the combined effects of water and carbon dioxide molecules. Since Earth's atmosphere has only one ten-thousandth as many carbon dioxide molecules as the atmosphere of Venus does, our greenhouse effect pales in comparison. As we continue to add more carbon dioxide by burning fossil fuels, we steadily increase Earth's greenhouse effect, performing an

unintended global experiment to determine just what delete-
rious effects arise from the additional trapping of heat. On
Venus, the atmospheric greenhouse effect, produced entirely
by carbon dioxide molecules, raises the temperature by hundreds of degrees, giving Venus's surface furnacelike temperatures close to 500 degrees Celsius (900°F)—the hottest in the
solar system.

How did Venus reach this sorry state? Scientists apply the
apt term "runaway greenhouse effect" to describe what happened as the infrared radiation trapped by Venus's atmosphere
raised the temperatures and encouraged liquid water to evaporate. The additional water in the atmosphere trapped infrared
even more effectively, increasing the greenhouse effect; this in
turn caused even more water to enter the atmosphere, ratch-
eting up the greenhouse effect still further. Near the top of
Venus's atmosphere, solar UV radiation would break the water
molecules apart into hydrogen and oxygen atoms. Because of
the high temperatures, the hydrogen atoms would escape,
while the heavier oxygen combined with other atoms, never
to form water again. With the passage of time, all the water
that Venus once had on or near its surface has been essentially
baked out of the atmosphere and lost to the planet forever.

Similar processes occur on Earth, but at a much lower rate
because we have much lower atmospheric temperatures. Our
mighty oceans now cover most of Earth's surface, though their
modest depth gives them only about one five-thousandth of
Earth's total mass. Even this small fraction of the total allows
the oceans to weigh in at a hefty 1.5 quintillion tons, 2 percent of which is frozen at any given time. If Earth should ever
undergo a runaway greenhouse effect like the one that has
occurred on Venus, our atmosphere would trap larger amounts

of solar energy, raising the air temperature and making the oceans evaporate swiftly into the atmosphere as they sustained a rolling boil. This would be bad news. Apart from the obvious ways that Earth's flora and fauna would die, an especially pressing cause of death would result from Earth's atmosphere growing 300 times more massive as it thickened with water vapor. We would be crushed and baked by the air we breathe.

———

Our planetary fascination (and ignorance) blossoms into true magnificence when we turn our attention from Venus to Mars. With its long-dry, still-preserved meandering riverbeds, floodplains, river deltas, networks of tributaries, and river-eroded canyons, Mars must have been a primeval Eden, with copious water flowing over its surface, billions of years ago. When we seek an orb in the solar system other than Earth that once boasted a flourishing water supply, we turn to Mars. For reasons unknown, however, today Mars has lost almost all of this water, so it now exhibits a bone-dry surface, with some evidence that tiny amounts of surface water exist briefly, in addition to a few lakes sheltered beneath its polar caps. A close examination of Venus and Mars, our sister and brother planets, forces us to look at Earth anew and to wonder how fragile our surface supply of liquid water may turn out to be.

Early in the twentieth century, imaginative observations of Mars by the noted American astronomer Percival Lowell led him to suppose that colonies of resourceful Martians had built an elaborate network of canals in order to redistribute water from Mars's polar ice caps to the more populated middle latitudes. To explain what he thought he saw, Lowell imagined a dying civilization that was exhausting its supply of water, like the city of Phoenix discovering that the Colorado River has its limits. In his thorough yet curiously misguided treatise entitled *Mars as the Abode of Life*, published in 1908, Lowell lamented the imminent end of the Martian civilization that he imagined he saw.

Indeed, Mars seems certain to dry up to the point that its surface can support no life at all. Slowly but surely, time will snuff life out, if it has not done so already. When the last living ember dies away, the planet will roll on through space as a dead world, its evolutionary career forever ended.

Lowell happened to get one thing right. If Mars ever had a civilization (or any kind of life at all) that required water on the surface, it must have faced catastrophe, because at some unknown time in Martian history, and for some unknown reason, all the surface water did dry up, leading to the exact fate for life—though in the past, not the present—that Lowell described. What happened to the water that flowed abundantly over Mars's surface billions of years ago remains an outstanding mystery among planetary geologists. Mars does have some water ice in its polar caps, which consist mainly of frozen carbon dioxide ("dry ice"), and a tiny amount of water vapor in its atmosphere. Although the polar caps contain the only significant amounts of water now known to exist on Mars, their total content of ice falls far below the amount needed to explain the ancient records of flowing water on Mars's surface.

If most of Mars's ancient water did not evaporate into space, its most likely hiding place lies underground, with the water trapped in the planet's subsurface permafrost. The evidence? Large craters on the Martian surface are more likely than small craters to exhibit dried mud spills over their rims. If the permafrost lies deep underground, to reach it would require a large collision. The deposit of energy from such an impact would melt this subsurface ice upon contact, causing it to splash upward. Craters with this mud-spill signature are more common in the cold, polar latitudes—just where we might expect the permafrost layer to be closer to the Martian surface. According to optimistic estimates of the Martian permafrost's ice content, the melting of Mars's subsurface layers would release enough water to give Mars a planetwide ocean tens of meters deep. A thorough search for contemporary (or fossil) life on Mars must include a plan to search in many locations, especially below the Martian surface. So far as the chance of finding life on Mars is concerned, the great question to be resolved asks, Does liquid water now exist anywhere on Mars?

Part of the answer leaps from our knowledge of physics. No liquid water can exist for long on the Martian surface, because the low atmospheric pressure there, less than 1 percent of the pressure on the surface of Earth, does not allow it. As enthusiastic mountaineers know, water vaporizes at progressively lower temperatures as the atmospheric pressure decreases. At the summit of Mount Whitney, where the air pressure falls to half of its sea-level value, water boils not at 100 but at 75 degrees Celsius. On top of Mount Everest, with air pressure only a quarter of its sea-level value, boiling occurs at about 50 degrees. Twenty miles high, where the atmospheric pressure equals only 1 percent of what you feel on the sidewalks of New York, water boils at about 5 degrees Celsius. Soar a few miles higher, and liquid water will "boil" at 0 degrees—that is, it will vaporize as soon as you expose it to the air. Scientists use the word "sublimation" to describe the passage of a substance from solid to gas without any intervening liquid stage. We all know sublimation from our youth, when the ice cream man opened his magic door to reveal not only the delicacies inside but also the chunks of "dry ice" that kept them cold. Dry ice offered the ice cream man a great advantage over familiar water ice: it sublimates from solid to gas, leaving no messy liquid to clean up. An old detective-story conundrum describes the man who hanged himself by standing on a cake of dry ice until it sublimated, leaving him suspended by a noose, and the detectives without a clue (unless they carefully analyzed the atmosphere in the room) as to how he did it.

What happens to carbon dioxide on Earth's surface happens to water on the surface of Mars. No chance for permanent pools of liquid exists there, even though the temperature on a warm day of the Martian summer rises well above 0 degrees Celsius. This may seem to draw a sad veil over the prospects for life—until we realize that liquid water could exist beneath the surface. Future missions to Mars, intimately bound up with the possibility of finding ancient or even modern life on the red planet, will direct themselves toward regions where they can drill into the Martian surface in search of the elixir of life, or relics of ancient life in former pools and riverbeds.

Elixir though it may appear, water represents a deadly substance among the chemically illiterate, to be avoided sedulously. In 1997, Nathan Zohner, a 14-year-old student at Eagle Rock Junior High School in Idaho, conducted a now famous (among science popularizers) science-fair experiment to test antitechnology sentiments and associated chemical phobia. Zohner invited people to sign a petition that demanded either strict control or a total ban of dihydrogen monoxide. He listed some of the odious properties of this colorless and odorless substance:

- It is a major component in acid rain.

- It eventually dissolves almost anything it comes in contact with.

- It can kill if accidentally inhaled.

- It can cause severe burns in its gaseous state.

- It has been found in tumors of terminal cancer patients.

Forty-three out of 50 people approached by Zohner signed the petition, six were undecided, and one was a great supporter of the molecule and refused to sign. Yes, 86 percent of the passersby voted to ban dihydrogen monoxide (H_2O) from the environment.

Maybe that's what really happened to the water on Mars.

———

Venus, Earth, and Mars together provide an instructive tale about the pitfalls and payoffs of focusing on water (or possibly other solvents) as the key to life. When astrophysicists considered where they might find liquid water, they originally concentrated on planets that orbit at the proper distances from their host stars to maintain water in liquid form—not too close in and not too far out. Thus we begin with the tale of Goldilocks.

Once upon a time—somewhat more than 4 billion years ago—the formation of the solar system was nearly complete. Venus had formed sufficiently close to the Sun for the intense solar energy to vaporize what might have been its water supply. Mars formed so far away that its water supply became forever frozen. Only one planet, Earth, had a distance "just right" for water to remain a liquid, and whose surface would therefore become a haven for life. This region around the Sun where water can remain liquid came to be known as the habitable zone.

Goldilocks liked things "just right," too. One of the bowls of porridge in the Three Bears' cottage was too hot. Another was too cold. The third was just right, so she ate it. Upstairs, one bed was too hard. Another was too soft. The third was just right, so Goldilocks slept in it. When the Three Bears came home, they discovered not only missing porridge but also Goldilocks fast asleep in their bed. (Don't remember how the story ends, but it remains a mystery to us why the Three Bears— omnivorous and occupying the top of the food chain—did not eat Goldilocks instead.)

The relative habitability of Venus, Earth, and Mars would intrigue Goldilocks, though the actual history of these planets is somewhat more complicated than three bowls of porridge. Four billion years ago, leftover water-rich comets and mineral-rich asteroids were still pelting the planetary surfaces, although at a much lower rate than before. During this game of cosmic billiards, some planets had migrated inward from where they had formed, while others were kicked into larger orbits. And among the dozens of planets that had formed, some moved on unstable orbits and crashed into the Sun or Jupiter. Others were ejected from the solar system altogether. In the end, the few planets that remained had orbits that were "just right" to survive billions of years.

Earth settled into an orbit with an average distance of 93 million miles from the Sun. At this distance, Earth intersects a measly one two-billionth of the total energy radiated by the Sun. If you assume that Earth absorbs all the energy received from the Sun, then our home planet's average temperature should be about 280 K (45°F), which falls midway between winter and summer temperatures. At normal atmospheric pressures, water freezes at 273 K and boils at 373 K, so we are well positioned with respect to the Sun for nearly all of Earth's water to remain contentedly in its liquid state.

Not so fast. In science you can sometimes get the right answer for the wrong reasons. Earth actually absorbs only two-thirds of the energy that reaches it from the Sun. The rest is reflected back into space by Earth's surface (especially by the oceans) and by its clouds. If we factor this reflection into the equations, the average temperature for Earth drops to about 255 K, well below the freezing point of water. Something must be operating to raise our average temperature to something a little more comfortable.

But wait once more. All theories of stellar evolution tell us that 4 billion years ago, when life was forming out of Earth's primordial soup, the Sun was a third less luminous than it is today, which would have left Earth's average temperature even further below freezing. Perhaps Earth in the distant past was simply closer to the Sun. Once the early period of heavy bombardment had ended, however, no known mechanisms could have shifted stable orbits back and forth within the solar system. Perhaps the greenhouse effect from Earth's atmosphere was stronger in the past. We don't know for sure. What we do know is that habitable zones, as originally conceived, have only peripheral relevance to whether life may exist on a planet within them. This has become evident from the fact that we cannot explain Earth's history on the basis of a simple habitable-zone model, and even more from the realization that water or other solvents need not depend on the heat from a star to remain liquid.

If we restrict our search for life to systems that use water as their go-to liquid, we must admit that Mars apparently remained outside the Sun's habitable zone during most of the solar system's history. Even today, with the Sun more luminous than it was billions of years ago, Mars barely has surface temperatures that allow liquid water to exist. More directly, we can now observe the entire globe of Mars in exquisite detail, and find only hints and traces of tiny amounts of liquid water.

These facts compel the abandonment of all hope of finding water, and possibly life, in Martian ponds today. The past, however, provided conditions far more hospitable to life: geological evidence points toward abundant water on Mars a few billion years ago. Most notably, some of the ancient craters on Mars closely resemble counterparts on Earth, though most of our planet's still contain water, while Mars's emphatically do not. One of Mars's ancient repositories of water, a 30-mile crater named Jezero, has a fan-shaped, clay-rich ancient delta almost surely formed from debris carried by water flowing through it. As a result, Jezero became the target of NASA's most recent spacecraft sent to explore the red planet.

Someday, astronauts may land on Mars and traverse its surface while searching for clues of ancient life. For our robot explorers, "someday" is today. While the prospect of sending astronauts to Mars remains years in the future, humanity's automated prospectors continue to increase their numbers and abilities. By the time that astronauts are ready for a trip to Mars, robotic explorers may have proved capable of equaling their abilities (except for television appeal) for longer periods and at much lower cost. In the summer of 2020, during one of the favorable lineups that occur at 26-month intervals, three different countries launched their own craft. The *Hope* spacecraft, created by the United Arab Emirates, entered orbit around Mars with a suite of advanced imaging instruments, joining six other orbiting probes that actively examine the entire Martian surface. China's *Tianwen-1*, also a planetary first by that country, included both an orbiter and *Zhurong*, a quarter-ton rover, that made a three-month examination of the landing site.

NASA, building on 45 years of experience that landed seven spacecraft on Mars, launched *Perseverance*, a one-ton rover with 19 cameras, a drill for obtaining soil samples, a laser and a fluorescence spectrometer to determine chemical composition, and a specialized instrument to look for "biosignature" molecules typical of living organisms on Earth. *Perseverance* also carried *Ingenuity*, the first helicopter on another planet, to test the feasibility of drone flights in an atmosphere with only 1 percent of Earth's thickness, helped by the fact that Mars has a surface gravity only 38 percent of Earth's.

Looking toward the ever-unknown future, NASA provided *Perseverance* with 39 hypersterilized tubes to contain soil samples obtained with the spacecraft's drill. These tubes will be left at one or more well-marked locations for future explorers— human or automated—to retrieve and to return to Earth for detailed analysis. The best hopes for finding microscopic life on Mars, if it exists, resides in bringing samples of Mars back to our planet's laboratories, rather than in landing comparatively primitive laboratories on the Martian surface.

———

Beyond Mars, solar energy continues to diminish, leaving planets and their moons in a comparative deep freeze that once seemed to make water-based life difficult, if not impossible. But during the past decades, ever-more-capable spacecraft sent to study the major objects beyond Mars have identified (so far!) five different objects that could support life. Since none of these objects orbit within the Sun's habitable zone, this fact demonstrates that what seem natural assumptions may prove both false and unnecessarily limiting.

The closest of these five objects offers the most recent, and perhaps the most surprising, example, one that has rarely, if ever, been previously mentioned in search-for-life circles. Ceres, the largest of the asteroids, has a diameter close to 600 miles, one-quarter of the Moon's, and orbits the Sun at two and three-quarters times the Earth-Sun distance. In 2015, after visiting Mars and the asteroid Vesta, NASA's *Dawn* spacecraft used its innovative ion-thruster propulsion system to achieve an orbit around Ceres, and eventually to shrink its orbit to pass only 20 miles above the asteroid's surface. After *Dawn* exhausted its propellant in 2018, NASA left the spacecraft to eternal silence in an orbit around Ceres, but continued to fund scientists at the Jet Propulsion Laboratory to interpret the years of data that *Dawn* had provided. Late in 2020, their efforts brought forth a startling announcement: abundant underground water exists on Ceres.

This conclusion arose from a variety of *Dawn*-made observations. Images showed a large crater named Occator, judged to be only about 20 million years old; detailed measurements of Ceres's gravitational field revealed the density of subsurface material; and spectroscopic observations demonstrated that Ceres has a surface covered with recently deposited material, including a water-and-salt compound called hydrohalite. On Earth, hydrohalite occurs abundantly in Arctic and Antarctic ice, but its appearance on Ceres marks an extraterrestrial first.

Ceres's surface temperature of 235 K (−38°C) implies that any liquid water must reside underground. *Dawn's* measurements suggest that below Occator crater are sizable deposits of brine (heavily salted water, with a freezing point many degrees below 0°C), rich in hydrohalite (which freezes only at −30°C), and loaded with grit—rocks of tiny size. In other words, we can expect to find plenty of salty cold mud beneath the asteroid's surface.

Ceres possesses a key weapon in its (imagined) struggle to maintain some sort of liquid: clathrates! This exotic name refers to a sort of "ice-nine" (from Kurt Vonnegut Jr.'s novel *Cat's Cradle*). Clathrates look much like ordinary water ice, but they have much more complexity, in which cages of water molecules surround small molecules of gas. This structure inhibits the flow of heat while making clathrate ice hundreds of times stronger than water ice. On Earth, the cages provided by clathrate molecules perform an existential role for us by trapping methane within their lattice-like structures: Without clathrates, underground methane molecules would flood our atmosphere, dooming humanity with their heat-trapping greenhouse properties.

Dawn's scientific team concluded that the existence of liquid, or at least of oozy mud, beneath Ceres's surface implies that the asteroid's surface has plenty of clathrates. We shall meet these intriguing entities again, much farther out in the solar system. Meanwhile, Ceres-oriented astrophysicists hope for either a spacecraft that could land and then hop from place to place to drill into the surface to search for the likeliest indications of life, or one that could land once, secure a soil sample, and return it to Earth.

Ceres may demonstrate the attraction of mud, but Jupiter's moon Europa offers the real deal: worldwide water. This satellite, about the same size as our Moon, shows crisscrossing cracks on its surface that change on time scales of weeks or months. To expert geologists and planetary scientists, this behavior implies that Europa has a surface made almost entirely of water ice, like a giant Antarctic ice sheet girdling an entire world. And the changing appearance of the rifts and rills in this icy surface leads to a startling conclusion: the ice apparently floats on a worldwide ocean. Only by invoking liquid beneath the icy surface can scientists satisfactorily explain what they have seen thanks to the stunning successes of the *Voyager* and *Galileo* spacecraft. Since we observe changes on the surface all around Europa, we may conclude that a worldwide ocean of liquid must underlie that surface.

What liquid could this be, and why should that substance remain liquid? Impressively, planetary scientists have reached two fairly firm additional conclusions: the liquid is water, and it remains liquid because of tidal effects on Europa produced by the giant planet Jupiter and other surrounding moons. The fact that water molecules are more abundant than ammonia, ethane, or methyl alcohol makes it the likeliest substance to provide the liquid beneath Europa's ice, and the existence of this frozen water likewise implies that more water exists in the immediate neighborhood.

But how can water remain a liquid, when the solar-induced temperatures in Jupiter's vicinity are only about 120 K (−150°C)? Europa's rocky interior remains relatively warm because of the threefold interplay that gravitational forces from Jupiter and the two large moons nearby, Io and Ganymede, exert upon it. At all times, the parts of Io and Europa closest to Jupiter feel a stronger force of gravity from the giant planet than the parts farthest away. These differences in force slightly elongate the solid moons in the direction facing Jupiter. But as the moons' distances from Jupiter change during their slightly elliptical orbits, Jupiter's tidal effect—the difference in force exerted on the moons' near and far sides—also changes, producing small pulses in their already distorted shapes. The modest gravitational forces from Io and Ganymede prevent Europa from achieving a circular orbit, thereby maintaining the continuously changing distortion that heats the moon's interior. Like a squash or racquet ball continually deformed by impact, any system that undergoes continuing structural stress will have its internal temperature rise.

———————

With a distance from the Sun that would otherwise guarantee a forever-frozen ice world, Io's stress level earns it the title of the most geologically active place in the entire solar system—complete with belching volcanoes, surface fissures, and plate tectonics. Some have analogized modern-day Io to the early Earth, when our planet was still piping hot from its episode of formation. Inside Io, the temperature rises to the point that volcanoes continually blast evil-smelling compounds of sulfur and sodium many miles above the satellite's surface. Io in fact has too high a temperature for liquid water to survive, but Europa, which undergoes less tidal flexing than Io because it is farther from Jupiter, heats more modestly, though still significantly. In addition, Europa's worldwide ice cap puts a pressure lid on the liquid below, preventing the water from evaporating and allowing it to exist for billions of years without freezing.

So far as we can tell, Europa was born with its water ocean below and its ice above, and has maintained that ocean, close to the freezing point but still above it, through 4.5 billion years of cosmic history.

Astrobiologists therefore view Europa's worldwide ocean as a prime target for investigation. Within a few years, we may have a mission for close study of this fascinating world. *Juno*, the NASA spacecraft that followed *Galileo* into orbit around Jupiter, studied the planet from 2016 to 2021. The mission aimed not to search for life on Europa but instead to study Jupiter itself, seeking clues to the formation of the Sun's largest planet by measuring the mass of its core, the strength of its magnetic field, and the composition of its thick atmosphere. In compensation, the next mission to Jupiter, NASA's *Europa Clipper*, will likewise perform dozens of highly elongated orbits around Jupiter, but these will be designed for close-up study of Europa, at times passing only 15 miles above the moon's surface. In addition to cameras to obtain detailed images and spectrometers to determine Europa's exterior composition, the spacecraft will carry instruments designed to probe below the moon's surface. Measurements of Europa's magnetic field should reveal the depth and salinity of subsurface liquid, which can be mapped with ice-penetrating radar and, less directly, with precise gravitational measurements that distinguish rock from liquid. The most exciting results could come from plumes of water vapor, seen by the Hubble Space Telescope in 2012. If *Europa Clipper* observes such plumes, it could measure their chemical composition and thus determine the conditions in the subsurface ocean. These in turn could reveal the likelihood that life swims in Europa's ocean or crawls along the ocean floor.

In view of the fecundity of life within our own oceans, Europa remains the most tantalizing place in the solar system to search for life beyond Earth. The discovery of enormous masses of organisms at depths a mile or more beneath the basalts of Washington State, living mainly on geothermal heat, suggests that we may someday find the Europan oceans alive with organisms unlike any on our planet, presumably primitive in design because of their limited supplies of energy. But just imagine going ice fishing on Europa! Only one pressing question would remain: Would we call any such creatures Europans or Europeans?

Mars and Europa currently offer targets numbers one and two in the search for extraterrestrial life within the solar system; Ceres, a complete latecomer, has only begun to gather its supporters. But two more enormous "Search Me" signs appear, on two of Saturn's moons, Titan and Enceladus, which together orbit the Sun at almost twice the distance of Jupiter and its moons.

Saturn's one giant moon, Titan, ties Jupiter's champion, Ganymede, as the largest moon in the solar system. Half again as large as our own Moon, Titan possesses a thick atmosphere, a property unequaled by any other moon, or by the planet Mercury, which is almost as large as Titan but much closer to the Sun, whose heat forces any Mercurian gases to escape into space. Unlike the atmospheres of Mars and Venus, Titan's atmosphere, many dozen times thicker than Mars's, consists primarily of nitrogen molecules, just as Earth's does. Floating within this transparent nitrogen gas, enormous numbers of aerosol particles form a permanent Titanian smog that forever shrouds the moon's surface from our gaze. As a result, speculation about life's possibilities has enjoyed a field day on Titan. Titan's surface temperature, close to 94 K (–179°C), falls far below those that allow liquid water to exist, but provides just the right temperature for liquid ethane, a carbon-hydrogen compound familiar to those who refine petroleum products. For decades, astrobiologists have imagined ethane lakes on Titan, chockful of organisms that float, eat, meet, and reproduce. In 2005, the first tentative investigation of the giant moon replaced speculation with exploration.

———

In October 1997, the *Cassini-Huygens* mission to Saturn, a collaboration of NASA and the European Space Agency (ESA), left Earth on a seven-year journey, receiving two gravity boosts from Venus, another from Earth, and one more from Jupiter before it reached Saturn, where it fired rockets to achieve an orbit around the ringed planet. Soon afterward, the *Huygens* probe, built by the ESA, detached itself from the *Cassini* spacecraft and descended through Titan's opaque clouds, deploying a heat shield to prevent frictional burning from its rapid passage through the upper atmosphere and a series of parachutes to slow down in the lower atmosphere. Safely on the satellite's surface, *Huygens* used six instruments to measure the temperature, density, and chemical composition of Titan's atmosphere, sending its data and images back to Earth via the *Cassini* spacecraft. The weight limits on its batteries allowed *Huygens* only a few hours of life on Titan, sufficient to secure dozens of images of a surface strewn with rocks and pebbles.

Repeatedly guided into new orbits around the ringed planet by its masters on Earth, *Cassini* spent a dozen years in close examination of Saturn and its moons, including six new ones discovered by the mission. Onboard radar pierced Titan's smog for a global view of the moon, and revealed numerous lakes (one of them larger than any of our Great Lakes), made not of liquid water, since the moon's temperature remains close to −179 degrees Celsius, but rather of liquid methane and ethane, compounds of hydrogen and carbon within which, at least in theory, primitive forms of life might have arisen. Titan-oriented astrophysicists judge that clathrates probably add strength to Titan's surface, allowing the ethane/methane lakes to endure more readily. Their claim awaits verification by future spacecraft expeditions.

———

Among Saturn's six varied, medium-sized moons, which have diameters ranging from 250 to 950 miles, by far the most interesting in the search for life is Enceladus. With only 15 percent of our Moon's or Europa's diameter, and 10 percent of Titan's, this distant world has an ice-coated surface, with a lack of craters that testifies to recent resurfacing, a process in which geysers release high-speed streams of gases that include some of the molecules that provide building blocks for life. *Cassini*'s measurements of Enceladus's gravity field demonstrated that an ocean of liquid water exists below the surface, maintained as a liquid not through tidal flexing, like that which takes place within Europa, but rather thanks to heat released by radioactive rocks within the moon's core.

With this discovery, Enceladus joined Mars, Europa, Titan, and the much more recent addition, Ceres, on the list of prime targets for scientists who are attempting to find life on other solar-system objects. Despite Europa's much larger size, and the fact that Enceladus lies almost as far from Earth as Europa, astrophysicists dream of sending to both Europa and Enceladus probes capable of piercing their solid ice crusts to explore the murky waters beneath, where abundant life may be swimming in near darkness reminiscent of the deep waters on our planet. Turning this dream into reality will require many years, not least because astrophysicists estimate the thickness of the icy crusts surrounding these faraway worlds at 10 miles or so, although Enceladus's crust may be much thinner near its poles.

———————

Beyond the planetary realm of our system lies Pluto, added to the list of planets upon its discovery in 1930 but removed by force of its own properties and by a vote of the International Astronomical Union (IAU) in 2006. Before that demotion, NASA had secured funding to send the *New Horizons* spacecraft to Pluto, whose tiny size and immense distance from Earth had prevented even the Hubble Space Telescope from obtaining a detailed view. Launched shortly before the IAU's vote, *New Horizons* received a gravity boost from Jupiter and reached Pluto, after a nearly decade-long journey, in July 2015. Separated from its primary target by 40 times the distance between Earth and the Sun, the spacecraft could only fly by, but the images and other data that it sent home fulfilled all of the hopes and plans that it embodied.

Detailed images of Pluto and its large satellite Charon revealed unexpectedly complex surfaces, with noticeable differences between the two bodies. Pluto, with just over two-thirds of the Moon's diameter and one-sixth of its mass, has a satellite with a bit more than half of its own diameter. The two objects circle their common center of mass every 6.4 days, locked by their mutual attraction to the point that they keep the same faces pointed toward each other, as our Moon does toward Earth. Both of them have surface temperatures of about 53 K (−220°C). From our search-for-life viewpoint, we might reasonably expect that these impressively low surface temperatures rule out any form of liquid. *New Horizons*, however, saw mountains near Pluto's south pole, two and a half miles high, whose configuration suggests that they could be "cryovolcanoes," which eject not molten rock but, rather, frozen volatile molecules such as water, methane, and ammonia. Titan likewise has tall mountains thought likely to be cryovolcanoes; in both cases, the clathrates we met on Ceres probably provide crucial strength to structures such as these. The existence of cryovolcanoes suggests that beneath its surface, to which the word "icy" fails to do full justice, Pluto has a heat source, possibly the original heat created by the collisions that shaped the Pluto-Charon system. If so, future explorers could someday investigate whether, in a variant of the situation beneath the surfaces of Europa and Enceladus, liquid or semiliquid material has come to harbor strange forms of life.

———

Is Pluto a planet? The great trove of images and other data that *New Horizons* returned from its journey past Pluto and Charon has settled a modest controversy while, at least arguably, intensifying another. The remarkable variety of topography on Pluto and its major satellite, along with the impressive differences between two celestial objects in close proximity, eliminated the possibility that they might turn out to be no more than celestial rock piles, similar to the largest asteroids.

Instead their surfaces—especially Pluto's—show complex terrain that has changed over billions of years.

Perhaps inspired by these fascinating landscapes, public sentiment in favor of maintaining Pluto's status as a planet has remained strong. Those in the pro-Pluto camp have a champion in Alan Stern, the principal investigator for the *New Horizons* spacecraft. Stern and those who sympathize with him reject the 2006 verdict of the IAU. Those who long ago memorized the names of the Sun's planets from Mercury to Pluto refuse to accept this change, on occasion flashing their special sign: nine fingers held vertically, to show their support for Pluto the planet.

What was the IAU thinking? Without plunging into the complex definitions embodied in its resolution, a simplified analysis stresses two facts: First, Pluto is small, with barely one-fifth the mass of our own Moon. Second, and probably more significant, astrophysicists have now discovered an array of objects similar to Pluto in size, orbiting the Sun at distances significantly greater than Pluto's. The six largest of these "Trans-Neptunian Objects," or TNOs, are Pluto, Eris, Haumea, Makemake, Gonggong, and Quaoar. Eris has a diameter about 2 percent smaller than Pluto's, but its higher density gives it a mass 27 percent greater. Even Quaoar, number six in size on the list, has half Pluto's diameter.

If Pluto qualifies as a planet, Eris should share that rank. Haumea, with two-thirds of their diameters, has a good claim as well, and we may proceed through the list with no obvious point that would demarcate true planets from dwarf planets. In addition, the largest asteroids, starting with Ceres and its underground water, could submit their own claims. In Pluto's favor, some could argue that Charon, a comparatively immense satellite with a diameter greater than Quaoar's, marks Pluto as unique among the TNOs. But most astrophysicists have bitten the trans-Neptunian bullet and have chosen to designate all these larger TNOs as dwarf planets.

Meanwhile, Pluto abides, totally unconcerned, as perhaps we on Earth should also be, about whether we call it a true planet or a dwarf planet. Instead, we would do well to celebrate the fact that despite everything that seemed to point toward the opposite conclusion, Pluto may yet turn out to be life's most distant outpost in the solar system.

Whether or not we require water for life, must we restrict ourselves to planets, dwarf planets, and their moons, on whose solid surfaces (or beneath them) water and other liquids can accumulate in quantity? Not at all. Water molecules, along with several other household chemicals such as ammonia and methane and ethyl alcohol, appear routinely in cool interstellar gas clouds. Under special conditions of low temperature and high density, an ensemble of water molecules can be induced to transform and to funnel energy from a nearby star into an amplified, high-intensity beam of microwaves. The atomic physics of this phenomenon resembles what a laser does with visible light. But in this case, the relevant acronym is maser, for "microwave amplification by the stimulated emission of radiation." Not only does water occur practically everywhere in the galaxy, it also occasionally envelops you as well. The great problem faced by would-be life in interstellar clouds arises not from a lack of raw materials but from the extremely low densities of matter, which enormously reduce the rate at which particles collide and interact. If life takes millions of years to arise on a dense planet such as Earth, it might take trillions of years to do so at much lower densities—far more time than the universe has so far provided.

By completing our search for life in the solar system, we seem to have finished our tour of the fundamental questions linked to our cosmic origins. We cannot, however, leave this arena without a look at the great origin issue that lies in the future: the potential origin of our contact with other civilizations. No astronomical topic grips the public imagination more vividly, and none offers a better chance to draw together the strands of what we have learned about the universe. Now that we know something about how life might begin on other worlds, let's examine the chances of satisfying a human desire as deep as any, the wish to find other beings in the cosmos with whom we might talk things over.

16

SEARCHING FOR LIFE IN THE MILKY WAY GALAXY

We have seen that within our own solar system, Mars, Ceres, Europa, Titan, and Enceladus offer the best hopes for discovering extraterrestrial life, either alive or in fossil form. These five objects present by far the best chances for finding water or another substance capable of providing a liquid solvent within which molecules can meet to carry on life's work. Because these five objects seem most likely to have either above-ground ponds or underground oceans, most astrobiologists limit their hopes of finding life in the solar system to the discovery of primitive forms of life on one or more of them. Pessimists have a reasonable argument, someday to be upheld or refuted by actual exploration, that even though we may well find conditions suitable for life on one or more of this favored fivesome, life itself may well prove entirely absent. Either way, the results of our searches in coming decades on Mars, Ceres, Europa, Titan, and Enceladus will prove deeply significant in judging the prevalence of life in the cosmos. Optimists and pessimists already agree on one conclusion: if we hope to find advanced forms of life—life that consists of creatures larger than the simple, single-celled organisms that appeared first and remain dominant in Earth-life—then we must look far beyond the solar system, to planets that orbit stars other than the Sun.

Once upon a time, we could only speculate about the existence of these planets. Now that we have found 5,000 exoplanets, we may confidently predict that only time and more precise observations separate us from the discovery of Earth-like planets. The early years of the twenty-first century seem likely to mark the moment in history when we acquired real evidence for an abundance of habitable worlds throughout the cosmos. Thus the first two terms in the Drake equation, which together measure the numbers of planets orbiting stars that last for billions of years, now imply high rather than low values. The next two terms, however, which describe the probability of finding planets suitable for life, and of life actually springing into existence on such planets, remain nearly as uncertain as they did before the discovery of exoplanets. Even so, our attempts to estimate these probabilities seem to rest on firmer grounds than our numbers for the final two terms: the probability that life on another world will evolve to produce an intelligent civilization, and the ratio of the average amount of time that such a civilization will survive to the lifetime of the Milky Way galaxy.

———

For the first five terms in the Drake equation, we can offer our planetary system and ourselves as a representative example, though we must always invoke the Copernican principle to avoid measuring the cosmos against ourselves, rather than the reverse. When we get to the equation's final term, however, and attempt to estimate the average lifetime of a civilization once it has acquired the technological capacity to send signals across interstellar distances, we fail to reach an answer even if we take Earth as a guide, since we have yet to determine how long our own civilization will last. We have now possessed interstellar-signaling capacity for nearly a century, ever since powerful radio transmitters began to send messages across Earth's oceans. Whether we last as a civilization for the next century, through the next millennium, or throughout a thousand centuries depends on factors far beyond our capacity to foresee, though many of the signs seem unfavorable to our long-term survival.

Asking whether our own fate corresponds to the average in the Milky Way takes us into another dimension of speculation. The final term in the Drake equation, which affects the result as directly as all the others, may therefore be judged as just plain unknown. If, in an optimistic assessment, most planetary systems contain at least one object suitable for life, and if life originates on a sensibly high fraction (say one-tenth) of those suitable objects, and if intelligent civilizations likewise appear on, perhaps, one-tenth of the objects with life, then at some point in the history of the Milky Way's 100 billion stars, 1 billion locations could produce an intelligent civilization. This enormous number springs, of course, from the fact that our galaxy contains so many stars, most of them much like our Sun. For a pessimistic view of the situation, simply change each of the numbers to which we assigned values from one-tenth to one chance in 10,000. Then the billion locations become 1,000, lower by a factor of 1 million.

This makes a major difference. Suppose that an average civilization, qualifying as a civilization by possessing interstellar communications ability, lasts for 10,000 years—

approximately one part in a million of the Milky Way's lifetime. On the optimistic view, a billion places give birth to a civilization at some point in history, so at any representative time, about 1,000 civilizations should be flourishing. The pessimistic view, in contrast, implies that in each representative era about 0.001 civilizations should exist, making ourselves a lone and lonely blip that temporarily rises high above the average value.

Which estimate has the greater chance of coming close to the true value? In science, nothing convinces so well as experimental evidence. If we hope to determine the average number of civilizations in the Milky Way, the best scientific approach would be to measure how many civilizations now exist. The most direct way to perform that feat would survey the entire galaxy, as the cast of television's *Star Trek* love to do, noting the number and type of each civilization that we encounter, if indeed we find any. (The possibility of an alien-free galaxy makes for boring television, rarely appearing on the small screen.) Unfortunately, this survey lies far outside our current technological capability and budgetary constraints.

Besides, surveying the entire galaxy would take millions of years, if not longer. Consider what a television program about interstellar-space surveys would be like if it limited itself by what we know of physical reality. Titled *Are We There Yet?*, a typical hour would show the crew complaining and bickering, aware that they had come so far yet still had so far to go. "We've read all the magazines," one of them might remark. "We're sick of each other, and you, Captain, are a great pain in the plethora." Then, while other crew members sang songs to themselves and still others entered private worlds of madness, a trailing long shot would remind us that the distances to other stars in the Milky Way are millions of times greater than the distances to other planets in the solar system.

Actually, this ratio describes only the distances to the Sun's closer neighbors, already so distant that their light takes many years to reach us. A full tour of the Milky Way would take us nearly 10,000 times farther. Hollywood films depicting interstellar spaceflight deal with this all-important issue by ignoring it (*Invasion of the Body Snatchers*, 1956), assuming that better rockets or improved understanding of physics will deal with it (*Star Wars*, 1977), or offering intriguing work-arounds such as freezing astronauts so that they can survive immensely long journeys (*Planet of the Apes*, 1968).

All of these approaches have a certain appeal, and some offer creative possibilities. We may indeed improve our rockets, which can now reach speeds of only about one ten-thousandth of the speed of light, which is the fastest we can hope to travel according to our current knowledge of physics. Even at the speed of light, however, travel to the nearest stars will take many years, and travel across the Milky Way close to a thousand centuries. Freeze-drying astronauts has some promise, but so long as those on Earth, who presumably will pay for the trip, continue unfrozen, the long passages of time before the astronauts return argues against easy funding. Given our short attention spans, by far the better approach to establishing contact with extraterrestrial civilizations—provided that they exist—appears right here on Earth. All we need do is to wait for them to contact us. This costs far less and can offer the immediate rewards that our society so eagerly craves.

Only one difficulty arises: Why should they? Just what about our planet makes us special to the point that we merit attention from extraterrestrial societies, assuming that they exist? On this point more than any other, humans have consistently violated the Copernican principle. Ask anyone why Earth deserves scrutiny, and you are likely to receive a sharp, angry stare. Almost all conceptions of alien visitors to Earth, as well as a sizable part of religious dogma, rest on the unspoken, obvious conclusion that our planet and our species rank so high on the list of universal marvels that no argument is needed to support the astronomically strange contention that our speck of dust, nearly lost in its Milky Way suburb, somehow stands out like a galactic beacon, not only demanding but also receiving attention on a cosmic scale.

This conclusion springs from the fact that the actual situation appears reversed when we view the cosmos from Earth. Then planetary matters bulk large, while the stars seem tiny points of light. From a quotidian point of view, this makes complete sense. Our success at survival and reproduction, like that of every other organism, has little to do with the cosmos that surrounds us. Among all astronomical objects, only the Sun and, to a much lesser extent, the Moon affect our lives, and their motions repeat with such regularity that they almost seem part of the Earthbound scene. Our human consciousness, formed on Earth from countless encounters with terrestrial creatures and events, understandably renders the extraterrestrial scene as a far-distant backdrop to the important action at center stage. Our error lies in assuming that the backdrop likewise regards ourselves as the center of activity.

Because each of us adopted this erroneous attitude long before our conscious minds attained any dominion or control over our patterns of thinking, we cannot eliminate it entirely from our approach to the cosmos even when we choose to do so. Those who impose the Copernican principle must remain ever vigilant against the murmurings of our reptilian brains, assuring us that we occupy the center of the universe, which naturally directs its attention our way.

When we turn to reports of extraterrestrial visitors to Earth, we must recognize another fallacy of human thought, as omnipresent and self-deceiving as our anti-Copernican prejudices. Human beings trust their memories far more than reality can justify. We do so for the same survival-value reasons that we regard Earth as the center of the cosmos. Memories record what we perceive, and we do well to pay attention to this record if we seek to draw conclusions for the future.

Now that we have better means of recording the past, however, we know better than to rely on individual memories for all matters of importance to society. We transcribe congressional debates and laws in print and in digital form, we record videos of crime scenes, and we make surreptitious audio recordings of criminal activity, because we recognize these media as superior to our own brains for creating a permanent record of past events.

One great apparent exception to this rule remains. We continue to accept eyewitness testimony as accurate, or at least probative, in legal proceedings. We do so despite test after test that demonstrates that each of us, despite our best intentions, will fail to remember events accurately, especially when those memories deal—as they usually do in cases important enough to go to trial—with unusual and exciting occurrences. Our legal system accepts eyewitness testimony because of its long tradition, its emotional resonance, and most of all because it often provides the only direct evidence of past events. Nevertheless, every courtroom cry of "That's the man who held the pistol!" must be weighed against the many demonstrated cases where that was not the man, despite the witness's sincere belief to the contrary.

If we bear these facts in mind when we analyze reports of unidentified flying objects (UFOs), we can immediately recognize an enormous potential for error. By definition, UFO sightings are bizarre occurrences, which cause observers to discriminate among familiar and unfamiliar objects on the rarely examined celestial backdrop, and typically require rapid conclusions about these objects before they quickly disappear. Add to this the psychic charge arising from the observer's belief in having witnessed a tremendously unusual event, and we could hardly find a better textbook example of a situation likely to generate an erroneous memory. Our brains will always attempt to fit unusual sightings within the framework of our previous sightings of other objects, an impediment to full understanding that can affect experienced pilots as much as, or even more than, the general public, thanks to the hours that they have spent interpreting whatever they see in the sky.

During the 1950s, astrophysicist J. Allen Hynek, then a leading Air Force consultant on UFOs, liked to highlight this issue by whipping a miniature camera from his pocket, insisting that if he ever saw a UFO, he would use the camera to obtain valid scientific evidence, because he knew that eyewitness testimony would not qualify. Unfortunately, improvements in technology since that time allow the creation of fake images and video recordings barely distinguishable from honest ones, so that Hynek's plan would no longer allow us to put our faith in photographic evidence supporting a UFO sighting. In fact, when we consider the interaction of memory's fragile power with the inventiveness of human con artists, we cannot easily devise a test to discriminate between fact and fancy for any individual UFO sighting. Nevertheless, we may note that recent decades have taken us from situations unlikely to involve camera-carrying observers into a new era, with almost all of us carrying high-resolution video cameras all the time. The number of UFO sightings by ordinary citizens hasn't budged.

In 2021, the U.S. government released a number of videos taken from military aircraft that showed UFOs, now renamed UAPs (unidentified aerial phenomena), moving in tandem with the aircraft and performing high-speed, high-acceleration maneuvers far beyond both our current technology (so far as has become public) and the tolerance of any human undergoing such acceleration. The UAPs had no visible control surfaces or exhausts. Without radar data, which could reliably establish the distance, speed, and physical reality of the images, we cannot confidently distinguish real objects from glitches arising from the complexity of advanced optical systems, together with optical illusions such as mirages, and misidentifications by military pilots. Proponents of different explanations for these UAPs—optical phenomena, hitherto-secret craft created by the United States or its enemies, or spacecraft from another civilization—all agree that we require more data to achieve a definitive explanation for the military videos.

When we turn to the more extreme phenomenon of UFO abductions, the ability of the human psyche to trump reality becomes even more apparent. Although hard numbers cannot be easily obtained, in the pre-smartphone era tens of thousands of people had come to believe that they had each been taken aboard an alien spacecraft and subjected to examinations, often of the most humiliating variety. From a calm perspective, stating this claim suffices to refute it as reality. Direct application of the principle of Occam's razor, which calls for the simplest explanation that fits the alleged facts, leads to the conclusion that these abductions have been imagined, not undergone. Because nearly all of the retellings place the abduction deep in the nighttime, and the majority in the midst of sleep, the likeliest explanation involves the hypnagogic state, the boundary between sleep and waking. For many people, this state brings visual and auditory hallucinations, and sometimes a "waking dream," in which the person feels conscious but unable to move. These effects pass through the filters of our brains to yield seemingly real memories, capable of arousing unshakable belief in their certainty.

Compare this explanation of UFO abductions with an alternative, that extraterrestrial visitors have singled out Earth and arrived in sufficient numbers to abduct humans by the thousands, though only briefly, and apparently to examine them closely (but should they not have long ago learned whatever they cared to—and could they not abduct sufficient corpses to learn human anatomy in detail?). Some stories imply that aliens extract some useful substances from their abductees, or plant their seeds into female victims, or alter the abductees' mind patterns to avoid later detection (but in that case could they not eliminate abduction memories entirely?). These assertions cannot be dismissed categorically, any more than we can rule out the possibility that alien visitors wrote these words, attempting to lull human readers into a false sense of security that will further the aliens' plans for world or cosmic domination. Instead, relying upon our ability to analyze situations rationally, and to discriminate between more likely and less likely explanations, we can assign an extremely low probability to the abduction hypothesis.

One conclusion seems unassailable by UFO skeptics and believers alike. If extraterrestrial societies do visit Earth, they must know that we have created worldwide capabilities for disseminating information and entertainment, if not for distinguishing one from the other. To say that these facilities would be open to any alien visitors caring to use them amounts to a gross understatement. They would receive immediate permission (come to think of it, they might not need it), and could make their presence felt in a minute—if they cared to. The absence of apparent extraterrestrials from our television screens testifies either to their absence from Earth or to their unwillingness to reveal themselves to our gaze—the "shyness" problem. The second explanation raises an intriguing conundrum. If alien visitors to Earth choose not to be detected, and if they possess technology far superior to ours, as their journeys across interstellar distances imply, why can they simply not succeed in their plans? Why should we expect to have any evidence—visual sightings, crop circles, pyramids built by ancient astronauts, memories of abductions—if the aliens prefer that we don't? They must be messing with our minds, enjoying their little game of cat and mouse. Quite probably they are secretly manipulating our leaders, too, a conclusion that snaps much of politics and entertainment into immediate focus.

The UFO phenomenon highlights an important aspect of our consciousness. Believing though we do that our planet forms the center of creation, and that our starry surroundings must decorate our world, rather than the reverse, we nevertheless maintain a strong desire to connect with the cosmos, manifested in mental activities as disparate as credence in extraterrestrial-visitor reports and belief in a benevolent deity that sends thunderbolts and emissaries to Earth. The roots of this attitude lie in the days when a self-evident distinction existed between the sky above and Earth below, between the objects we could touch and scratch and those that moved and shone but remained forever beyond our reach. From these differences we drew distinctions between the earthly body and the cosmic soul, the mundane and the marvelous, the natural and the supernatural. The need for a mental bridge connecting these two apparent aspects of reality has informed many of our attempts to create a coherent picture of our existence. Modern science's demonstration that we are stardust has thrown an enormous wrench into our mental equipment, from which we are still struggling to recover. UFOs suggest new messengers from the other part of existence, all-powerful visitors who well know what they are up to while we remain ignorant, barely aware that the truth is out there. This attitude was captured well in the classic film *The Day the Earth Stood Still* (1951), in which an alien visitor, far wiser than we, comes to Earth to warn that our violent behavior may lead to our own destruction.

Our innate feelings about the cosmos manifest a dark side that projects our feelings about human strangers onto nonhuman visitors. Many a UFO report contains phrases similar to "I heard something odd outside, so I took my rifle and went to see what it was." Films that depict aliens on Earth likewise slip easily into a hostile mode, from the Cold War epic *Earth vs. the Flying Saucers* (1956), in which the military blasts away at alien spacecraft without pausing to ask their intentions, to *Signs* (2002), in which the peace-loving hero, with no rifle at hand, uses a baseball bat to chastise his trespassers—

a method not likely to succeed against actual aliens capable of crossing interstellar distances.

The greatest arguments against interpreting UFO reports as evidence for extraterrestrial visitors reside in our planet's unimportance, together with the vast distances between the stars. Neither can be regarded as absolute bars to this interpretation, but in tandem they form a powerful argument. Must we, then, conclude that because Earth lacks popular appeal, our hopes of finding other civilizations must await the day when we can expend our own resources to embark on journeys to other planetary systems? Not at all. The scientific approach to establishing contact with other civilizations within the Milky Way and beyond, should they exist, has always relied on letting nature work in our favor. This principle redirects the question "What aspect of extraterrestrial civilizations would we find most exciting?" (answer: visitors in the flesh) into the scientifically fruitful one: What seems to be the most likely means of establishing contact with other civilizations? Nature, and the immense distances between stars, supply the answer—use the cheapest, fastest means of communication available, which presumably holds the same rank elsewhere in the galaxy.

The cheapest and fastest means of sending messages between the stars uses electromagnetic radiation, the same medium that carries almost all long-range communication on Earth. Radio waves have revolutionized human society by allowing us to send words and pictures around the world at 186,000 miles per second. These messages travel so rapidly that even if we beam them up to a geostationary satellite orbiting at an altitude of 23,000 miles, which relays them to another part of Earth's surface, they undergo a time delay on each leg of their journey much shorter than one second.

Over interstellar distances, the time lag grows longer, though it remains the shortest we can hope to achieve. If we plan to send a radio message to Alpha Centauri, the star system closest to the Sun, we must plan on a travel time of 4.4 years in each direction. Messages that travel for, say, 20 years can reach several hundred stars, or any planets that orbit them. Thus if we are prepared to wait for a round-trip of 40 years, we could beam a message toward each of these stars, and eventually find out whether we receive a reply from any of them. This approach assumes, of course, that if civilizations exist close to any of these stars, they have a command of radio, and an interest in its application, at least equal to ours.

The fundamental reason why we don't adopt this approach in searching for other civilizations lies not in its assumptions but in our attitudes. Forty years is a long time to wait for something that may never happen. (Yet if we had beamed out messages 40 years ago, by now we would have some serious information about the abundance of radio-using civilizations in our region of the Milky Way.) The only serious attempt in this direction occurred in the 1970s, when astrophysicists celebrated the upgrading of what was then the world's largest radio telescope, near Arecibo, Puerto Rico, by using it to beam a message for a few minutes in the direction of the star cluster M13. Since the cluster lies 25,000 light-years away, any return message will be a long time coming, rendering the exercise more a demonstration than an actual casting call. In case you think that discretion has inhibited our broadcasting (for it is good to be shifty in a new country), recall that all of our post–

World War II radio and television broadcasting, as well as our powerful radar beams, has sent spherical shells of radio waves into space. Expanding at the speed of light, the "messages" from the *Honeymooners* and *I Love Lucy* era have already washed over tens of thousands of stars, while *Hawaii Five-O* and *Charlie's Angels* have reached many hundreds. If other civilizations really could disentangle individual programs from the cacophony of Earth's radio emission—now comparable to or stronger than that from any solar-system object, including the Sun—there might be some truth to the playful speculation that the content of these programs explains why we have heard nothing from our neighbors: because they find our programming so appalling, they've concluded there is no sign of intelligent life on Earth.

A message might arrive tomorrow, laden with intriguing information and commentary. Herein lies the greatest appeal of communication by electromagnetic radiation. Not only is it cheap (sending 50 years of television broadcasts into space has cost less than a single spacecraft mission), it is also instantaneous—provided that we can receive and interpret another civilization's emission. This also provides a fundamental aspect of UFO excitement, but in this case we might actually receive transmissions that could be recorded, verified as real, and studied for as long as it would take to understand them.

In the search for extraterrestrial intelligence, shortened to SETI by the scientists who engage in it, the focus remains on searching for radio signals, though the alternative of looking for signals sent with light waves should not be rejected.

Although light waves from another civilization must compete with numerous natural sources of light, laser beams offer the opportunity to concentrate the light into a single color or frequency—the same approach that allows radio waves to carry messages from different radio or television stations. So far as radio waves go, our hopes for success in SETI rest with antennas that can survey the sky, receivers that record what the antennas detect, and powerful computers that analyze the receivers' signals in a search for the unnatural. Two basic possibilities exist: we might find another civilization by eavesdropping on its own communications, some of which leak into space in the same way that our radio and television broadcasts do; or we might discover deliberately beamed signals, meant to attract the attention of previously uncatalogued civilizations such as our own.

Eavesdropping clearly presents a more difficult task. A beamed signal concentrates its power in a particular direction, so detecting that signal becomes much easier if it is deliberately sent toward us, whereas signals that leak into space diffuse their power more or less evenly in all directions and are therefore much weaker at a particular distance from their source than a beamed signal. Furthermore, a beamed signal would presumably contain some easy warm-up exercises to tell its recipients how to interpret it, whereas radiation that leaks into space presumably carries no such user's manual. Our own civilization has leaked signals for many decades, and has sent a beamed signal in one particular direction for a few minutes. If civilizations are rare, any attempts to find them ought to concentrate on eavesdropping and avoid the lure of hoping for deliberately beamed signals.

With ever-better systems of antennas and receivers, SETI

proponents have begun to eavesdrop on the cosmos, hoping to find evidence for other civilizations. Precisely because we have no guarantee that we shall ever hear anything by eavesdropping, those who engage in these activities have had difficulty securing funding. In the early 1990s, the U.S. Congress supported a SETI program for a year, until competing forces pulled the plug. SETI scientists now draw their support from a variety of modest funding sources and employ innovative techniques, such as analyzing any data received by radio telescopes as they study cosmic objects for other purposes. Some funding has come from wealthy individuals, including Yuri Milner, in support of the "Breakthrough Listen" project, which will use radio and optical telescopes to study selected exoplanets identified by NASA's recently launched TESS satellite.

In searching for radio waves produced by an extraterrestrial civilization, a fundamental difficulty is searching through billions of possible frequencies at which other civilizations might be broadcasting. On Earth, we divide the radio spectrum into relatively wide bands, so that only a few hundred different frequencies exist for radio and television broadcasts. In principle, however, alien signals might be confined so narrowly in frequency that the SETI dial would need billions of entries. Powerful computer systems, which lie at the heart of current SETI efforts, can meet this challenge by analyzing hundreds of millions of frequencies simultaneously. On the other hand, they have not yet found anything suggestive of another civilization's radio communications.

Midway through the twentieth century, the Italian genius Enrico Fermi, perhaps the last great physicist to work both as an experimentalist and as a theorist, discussed extraterrestrial life during lunch with his colleagues. Agreeing that nothing particularly special distinguishes Earth as an abode for life, the scientists reached the conclusion that life ought to be abundant in the Milky Way. In that case, Fermi asked, in a query that ripples across the decades, *where are they?*

Fermi meant that if many places in our galaxy have seen the advent of technologically advanced civilizations, surely we should have heard from one of them by now, by radio or laser messages if not by actual visits. Even if most civilizations die out quickly, as ours may, the existence of large numbers of civilizations implies that some of them should have sufficiently extended lifetimes to mount long-term searches for others. Even if some of these long-lived civilizations do not care to engage in such searches, others will. So the fact that we have no scientifically verified visits to Earth, nor reliable demonstrations of signals produced by another civilization, may prove that we have badly overestimated the likelihood that intelligent civilizations arise in the Milky Way.

Fermi had a point. Every day that passes adds a bit more evidence that we may be alone in our galaxy. However, when we examine the actual numbers, the evidence looks weak. If several thousand civilizations exist in the galaxy at any representative time, the average separation between neighboring civilizations will be a few thousand light-years, a thousand times the distance to the closest stars. If one or more of these civilizations has lasted for millions of years, we might expect that by now they should have sent us a signal, or revealed themselves to our modest eavesdropping efforts. If, however, no civilization attains anything like this age, then we shall have to work harder to find our neighbors, because none of them may be engaged in a galaxywide attempt to find others, and none of them may be broadcasting so powerfully that our present eavesdropping efforts can find them.

Thus we remain in a familiar human condition, poised at the edge of events that may not occur. The most important news in human history could arrive tomorrow, next year, or never. Let us go forth into a new dawn, ready to embrace the cosmos as it surrounds us, and as it reveals itself, shining with energy and replete with mystery.

CODA

THE SEARCH FOR OURSELVES IN THE COSMOS

Equipped with his five senses, man explores the universe
around him and calls the adventure science.
—EDWIN P. HUBBLE, 1948

Human senses display an astonishing acuity and range of sensitivity. Our ears can record the thunderous launch of the space shuttle, yet they can also hear a male mosquito buzzing in the corner of a room. Our sense of touch allows us to feel the crush of a bowling ball dropped on our big toe, or to tell when a one-milligram bug crawls along our arm. Some people enjoy munching on habanero peppers, while sensitive tongues can identify the presence of food flavors at a few parts per million. And our eyes can register the bright sandy terrain on a sunny beach, yet have no trouble spotting a lone match, freshly lit hundreds of feet away, across a darkened auditorium. Our eyes also allow us to see across the room and across the universe. Without our vision, the science of astronomy would never have been born and our capacity to measure our place in the universe would have remained hopelessly stunted.

In combination, these senses allow us to decode the basics of our immediate environment, such as whether it's day or night, or when a creature is about to pounce. But little did anybody know, until the past few centuries, that our senses alone offer only a narrow window on the physical universe.

Some people boast of a sixth sense, professing to know or see things that others cannot. Fortune-tellers, mind readers, and mystics top the list of those who claim mysterious powers. In doing so, they instill widespread fascination in others. The questionable field of parapsychology rests on the expectation that at least some people actually harbor this talent.

In contrast, modern science wields dozens of senses. But scientists do not claim that these are the expression of special powers, just special hardware that converts the information gleaned by these extra senses into simple tables, charts, diagrams, or images that our five inborn senses can interpret.

With apologies to Edwin P. Hubble, his remark at the head of this chapter, while poignant and poetic, should instead have been:

> Equipped with our five senses, along with telescopes and microscopes and mass spectrometers and seismographs and magnetometers and particle detectors and accelerators and instruments that record radiation from the entire electromagnetic spectrum, as well as new detectors of gravitational radiation, we explore the universe around us and call the adventure science.

Think of how much richer the world would appear to us, and how much sooner we would have discovered the fundamental nature of the universe, if we were born with high-precision, tunable eyeballs. Dial up the radio-wave part of the spectrum and the daytime sky turns as dark as night, except for some choice directions. Our galaxy's center appears as one of the brightest spots on the sky, shining brightly behind some of the principal stars of the constellation Sagittarius. Tune in to microwaves and the entire universe glows with a remnant from the early universe, a wall of light that set forth on its journey to us 380,000 years after the big bang. Tune in to X-rays and you will immediately spot the locations of black holes with matter spiraling into them. Tune in to gamma rays and see titanic explosions bursting forth from random directions about once a day throughout the universe. Watch the effect of these explosions on the surrounding material as it heats up to produce X-rays, infrared, and visible light.

If we were born with magnetic detectors, the compass would never have been invented because no one would ever need one. Just tune in to Earth's magnetic field lines and the direction of magnetic North looms like Oz beyond the horizon. If we had spectrum analyzers within our retinas, we would not have to wonder what the atmosphere is made of. Simply by looking at it we would know whether or not it contains sufficient oxygen to sustain human life. And we would have learned thousands of years ago that the stars and nebulae in our galaxy contain the same chemical elements as those found here on Earth.

And if we were born with big, sensitive eyes and built-in Doppler motion detectors, we would have seen immediately, even as grunting troglodytes, that the entire universe is expanding—that all distant galaxies are receding from us.

If our eyes had the resolution of high-performance micro-scopes, nobody would have ever blamed the plague and other sicknesses on divine wrath. The bacteria and viruses that made you sick would have been in plain view as they crawled on your food or slid through open wounds in your skin. With simple experiments, you could easily tell which of these bugs were bad and which were good. And the carriers of postoperative-infection problems would have been identified and dealt with hundreds of years earlier.

If we could detect high-energy particles, we would spot radioactive substances from great distances. No Geiger counters necessary. You could even watch radon gas seep through the basement floor of your home and not have to pay somebody to tell you about it.

The honing of our five senses from birth through childhood allows us as adults to pass judgment on events and phenomena in our lives, declaring whether or not they "make sense." Problem is, hardly any scientific discoveries of the past century have flowed from the direct application of our senses. They came instead from the direct application of sense-transcendent mathematics and hardware. This simple fact explains why, to the average person, relativity, particle physics, and 11-dimensional string theory make no sense. Add to this list black holes, wormholes, and the big bang. Actually, these concepts don't make much sense to scientists either, until we have explored the universe for a long time with all of our senses that are technologically available. What eventually emerges is a newer and higher level of "uncommon sense" that enables scientists to think creatively and to pass judgment in the unfamiliar underworld of the atom or in the mind-bending domain of higher-dimensional space. The twentieth-century German physicist Max Planck made a similar observation about discovery of quantum mechanics: "Modern physics impresses us particularly with the truth of the old doctrine which teaches that there are realities existing apart from our sense-perceptions, and that there are problems and conflicts where these realities are of greater value for us than the richest treasures of the world of experience."

Each new way of knowing heralds a new window on the universe—a new detector to add to our growing list of non-biological senses. Whenever this happens, we achieve a new level of cosmic enlightenment, as though we were evolving into supersentient beings. Who could have imagined that our quest to decode the mysteries of the universe, armed with a collection of artificial senses, would grant us insight into ourselves? We embark on this quest not from a simple desire but from a mandate of our species to search for our place in the cosmos. The quest is old, not new, and has garnered the attention of thinkers great and small, across time and across culture. Exploring the cosmos reveals not only the majesty of the universe, but also our history and our role within it.

ACKNOWLEDGMENTS

For reading and rereading the manuscript, ensuring that we mean what we say and say what we mean, we are indebted to Robert Lupton of Princeton University. His tandem expertise in astrophysics and the English language allowed the book to reach several notches higher than we had otherwise imagined for it. June Fox likewise offered highly useful suggestions in presenting this material. We are also grateful to Sean Carroll of the California Institute of Technology, Julie Castillo-Rogez of the Jet Propulsion Laboratory, Lloyd Knox of UC Davis, Tobias Owen of the University of Hawaii, Adam Riess of Johns Hopkins University and the Space Telescope Science Institute, Steven Soter of the American Museum of Natural History, Larry Squire of UC San Diego, Michael Strauss of Princeton University, and Tom Levenson of the Massachusetts Institute of Technology for key suggestions that improved several parts of the book. At W. W. Norton, Helen Thomaides helped to smooth the flow and increase the explanatory power of our text.

For expressing confidence in the project from the beginning, we thank Betsy Lerner of Dunow, Carlson & Lerner Literary

Agency, who saw our manuscript not only as a book but also as an expression of deep interest in the cosmos, deserving the broadest possible audience with whom to share the love.

Major portions of Part 2 and scattered portions of Parts 1 and 3 were adapted from essays that first appeared in *Natural History* magazine by NDT. For this, he is grateful to Peter Brown, the magazine's former editor in chief, and especially to Avis Lang, the magazine's former senior editor, who continued to work heroically as learned literary shepherds to NDT's writing efforts.

The authors further recognize support from the Sloan Foundation in the writing and preparation of this book. We continue to admire their legacy of support for projects such as this.

—Neil deGrasse Tyson, *New York City*
—Donald Goldsmith, *Berkeley, California*
September 2022

GLOSSARY OF SELECTED TERMS

absolute (Kelvin) temperature scale: *Temperature* measured on a scale (denoted by K) on which water freezes at 273.16 K and boils at 373.16 K, with 0 K denoting absolute zero, the coldest theoretically attainable temperature.

acceleration: A change in an object's speed or direction of motion (or both).

accretion: An infall of matter that adds to the mass of an object.

accretion disk: Material surrounding a massive object, typically a *black hole*, that moves in orbit around it and slowly spirals inward.

AGN: Astronomical shorthand for a *galaxy* with an active *nucleus*, a modest way of describing galaxies whose central regions shine thousands, millions, or even billions of times more brightly than the central regions of a normal galaxy. AGNs have a generic similarity to *quasars*, but they are typically observed at distances less than that of quasars, hence later in their lives than quasars themselves.

amino acid: One of a class of relatively small *molecules*, made of 13 to 27 *atoms* of *carbon*, *nitrogen*, *hydrogen*, *oxygen*, and *sulfur*, which can link together in long chains to form *protein molecules*.

ammonia (NH_3): A type of molecule containing one *nitrogen* and three *hydrogen atoms*, and a potential *solvent* for extraterrestrial life.

Andromeda galaxy: The closest large *spiral galaxy* to the Milky Way, approximately 2.4 million *light-years* from our own galaxy.

antimatter: The complementary form of matter, made of *antiparticles* that have the same mass but opposite sign of *electric charge* as the particles that they complement.

antiparticle: The *antimatter* complement to a particle of ordinary matter.

apparent brightness: The brightness that an object appears to have as an observer measures it, hence a brightness that depends on the object's *luminosity* and its distance from the observer.

Archaea: Representatives of one of the three domains of life, thought to be the oldest forms of life on Earth. All Archaea are single-celled and thermophilic (capable of thriving at temperatures above 50°–70° Celsius).

asteroid: One of the objects, made primarily of rock or of rock and metal, that orbit the Sun, mainly between the orbits of Mars and Jupiter, and range in size from 1,000 kilometers in diameter down to objects about 100 meters across. Objects similar to asteroids but smaller in size are called *meteoroids*.

astronomer: One who studies the *universe*. Used more commonly in the past, at a time before spectra were obtained of cosmic objects.

astrophysicist: One who studies the *universe* using the full tool kit enabled by the known laws of physics. The preferred term for astronomers in modern times.

atom: The smallest electrically neutral unit of an *element*, consisting of a *nucleus* made of one or more *protons* and zero or more *neutrons*, around which orbit a number of *electrons* equal to the number of protons in the nucleus. This number determines the chemical characteristics of the atom.

Bacteria: One of the three domains of life on Earth (formerly known as prokaryotes), single-celled organisms with no well-defined *nucleus* that holds genetic material.

barred spiral galaxy: A *spiral galaxy* in which the distribution of stars and gas in the galaxy's central regions has an elongated, bar-like configuration.

big bang: The scientific description of the origin of the *universe*, premised on the hypothesis that the universe began in an explosion that brought space and matter into existence approximately 14 billion years ago. Today the universe continues to expand in all directions, everywhere, as the result of this explosion.

black hole: An object with such enormous *gravitational force* that nothing, not even light, can escape from within a specific distance from its center, called the object's *black hole radius*.

black hole radius: For any object with a mass M, measured in units of the Sun's mass, a distance equal to 3M kilometers, also called the object's *event horizon*.

blue shift: A shift to higher *frequencies* and shorter *wavelengths*, typically caused by the *Doppler effect*.

brown dwarf: An object with a composition similar to a star's, but with too little mass to become a star by initiating *nuclear fusion* in its core.

carbohydrate: A *molecule* made only of *carbon*, *hydrogen*, and *oxygen atoms*, typically with twice as many hydrogen as oxygen atoms.

carbon: The element that consists of *atoms* whose *nuclei* each have six *protons*, and whose different *isotopes* each have six, seven, or eight *neutrons*.

carbon dioxide (CO_2): A type of *molecule* containing one *carbon* and two *oxygen atoms*.

***Cassini-Huygens* spacecraft:** The spacecraft launched from Earth in 1997 that reached Saturn in July 2004, after which the *Cassini* orbiter surveyed Saturn and its moons and released the *Huygens* probe to descend to the surface of Titan, Saturn's largest satellite.

Celsius or Centigrade temperature scale: The *temperature* scale named for the Swedish astronomer Anders Celsius (1701–1744), who introduced it in 1742, according to which water freezes at 0 degrees and boils at 100 degrees.

catalyst: A substance that increases the rate at which specific reactions between *atoms* or *molecules* occur, without itself being consumed in these reactions.

CBR: See *cosmic background radiation*.

cell: A structural and functional unit found in all forms of life on Earth.

Ceres: The largest *asteroid*, found in 2020 to possess large amounts of water beneath its surface.

Charon: Pluto's largest satellite, with more than half of Pluto's diameter.

chromosome: A single *DNA* molecule, together with the *proteins* associated with that molecule, which stores genetic information in subunits called *genes* and can transmit that information when cells *replicate*.

civilization: For *SETI* activities, a group of beings with interstellar communications ability at least equal to our own on Earth.

COBE (COsmic Background Explorer) satellite: The satellite launched in 1989 that observed the *cosmic background radiation* and made the first detection of small differences in the amount of this radiation arriving from different directions on the sky.

comet: A fragment of primitive solar-system material, typically a "dirty snowball" made of ice, rock, dust, and frozen *carbon dioxide* (dry ice).

compound: A synonym for *molecule*.

constellation: A localized group of stars, as seen from Earth, named after an animal, planet, scientific instrument, or mythological character, which in rare cases actually describes the star pattern; one of 88 such groups in the sky.

cosmic background radiation (CBR): The sea of *photons* produced everywhere in the *universe* soon after the *big bang*, which still fills the universe and is now characterized by a *temperature* of 2.73 K.

cosmic tension: The conflict between different values of the *Hubble constant* obtained from different approaches to measuring that value.

cosmological constant: The constant introduced by Albert Einstein into his equation describing the overall behavior of the *universe*, which describes the amount of energy, now called *dark energy*, in every cubic centimeter of seemingly empty space.

cosmologist: An *astrophysicist* who specializes in the origin and large-scale structure of the *universe*.

cosmology: The study of the *universe* as a whole, and of its structure and evolution.

cosmos: Everything that exists; a synonym for *universe*.

cyanogen: Molecules of HCN, a likely key precursor of life on Earth.

dark energy: *Energy* that is invisible and undetectable by any direct measurement, whose amount depends on the size of the *cosmological constant*, and which tends to make space expand.

dark matter: Matter of unknown form that emits no *electromagnetic radiation*, that has been deduced, from the *gravitational forces* it exerts on visible matter, to make up the bulk of all matter in the universe.

decoupling: The era in the *universe's* history when *photons* first had too little energy to interact with *atoms*, so that for the first time atoms could form and endure without being broken apart by photon impacts.

DNA (deoxyribonucleic acid) molecule: A long, complex *molecule* consisting of two interlinking spiral strands, bound together by thousands of cross-links formed from small molecules. When DNA molecules divide and *replicate*, they split lengthwise, splitting each pair of small molecules that form their cross-links. Each half of the molecule then forms a new replica of the original molecule from smaller molecules that exist in the nearby environment.

Doppler effect: The change in *frequency, wavelength*, and *energy* observed for *photons* arriving from a source that has a relative velocity of approach or recession along an observer's line of sight to the source. These changes in frequency and wavelength are a general phenomenon that occurs with any type of wave motion. They do not depend on whether the source is moving or the observer is moving; what counts is the relative motion of the source with respect to the observer along the observer's line of sight.

Doppler shift: The fractional change in the *frequency, wavelength*, and *energy* produced by the *Doppler effect*.

double helix: The basic structural shape of *DNA molecules*.

Drake equation: The equation, first derived by the American astrophysicist Frank Drake, that summarizes our estimate of the number of *civilizations* with interstellar communications capability that exist now or at any representative time.

dry ice: Frozen *carbon dioxide* (CO_2).

dust cloud: Gas clouds in interstellar space that are cool enough for *atoms* to combine to form *molecules*, many of which themselves combine to form dust particles made of millions of atoms each.

dynamics: The study of the motion and the effect of *forces* on the interaction of objects. When applied to the motion of objects in the solar system and the universe, this is often called celestial mechanics.

eavesdropping: The technique of attempting to detect an extraterrestrial *civilization* by capturing some of the *radio* signals used for the civilization's internal communications.

eccentricity: A measure of the flatness of an *ellipse*, equal to the ratio of the distance between the two "foci" of the ellipse to its long axis.

eclipse: The partial or total obscuration of one celestial object by another, as seen by an observer when the objects appear almost or exactly in line with each other.

electric charge: An intrinsic property of *elementary particles*, which may be positive, zero, or negative; unlike signs of *electric charge* attract one another and like signs of electric charge repel one another through *electromagnetic forces*.

electromagnetic force: One of the four basic types of *forces*, acting between particles with *electric charge*, and diminishing in proportion to the square of the distance between the particles. Recent investigations have shown that these forces and *weak forces* are different aspects of a single *electroweak force*.

electromagnetic radiation: Streams of *photons* that carry energy away from a source of photons, not to be confused with *gravitational radiation*. Photons are classified as *gamma rays, X-rays, ultraviolet, visible light, infrared*, microwaves, or *radio* waves, depending on their *energies, frequencies*, and *wavelengths*.

electron: An *elementary particle* with one unit of negative *electric charge*, which in an *atom* orbits the atomic *nucleus*.

electroweak forces: The unified aspect of *electromagnetic forces* and *weak forces*, whose aspects appear quite different at relatively low energies but become unified when acting at enormous energies such as those typical of the earliest moments of the *universe*.

elements: The basic components of matter, classified by the number of *protons* in the *nucleus*. All ordinary matter in the *universe* is composed of 92 elements that range from the smallest *atom, hydrogen* (with one proton in its nucleus), to the largest naturally occurring element, uranium (with 92 protons in its nucleus). Elements heavier than uranium have been produced in laboratories.

elementary particle: A fundamental particle of nature, normally indivisible into other particles. *Protons* and *neutrons* are usually designated as elementary particles although they each consist of three particles called *quarks*.

ellipse: A closed curve defined by the fact that the sum of the distances from any point on the curve to two interior fixed points, called foci, has the same value.

elliptical galaxy: A galaxy with an ellipsoidal distribution of stars, containing almost no interstellar gas or dust, whose shape seems elliptical in a two-dimensional projection.

Enceladus: One of Saturn's medium-sized moons, whose icy surface, concealing a worldwide ocean, makes it a prime target in the search for extraterrestrial life.

energy: The capacity to do work; in physics, "work" is specified by a given amount of *force* acting through a specific distance.

energy of mass: The *energy* equivalent of a specific amount of mass equals the mass times the square of the speed of light.

energy of motion: See *kinetic energy*.

enzyme: A type of *molecule*, either a *protein* or *RNA*, that serves as a site at which molecules can interact in certain specific ways, and thus acts as a *catalyst*, increasing the rate at which particular molecular reactions occur.

escape velocity: For a projectile or spacecraft, the minimum speed required for an outbound object to leave its point of launching and never return to the object, despite the object's *gravitational force*.

Eukarya: The totality of organisms classified as *eukaryotes*.

eukaryote: An organism, either single-celled or multicellular, that keeps the genetic material in each of its cells within a membrane-bounded nucleus. Compare with *prokaryotes*.

Europa: One of Jupiter's four large satellites, notable for its icy surface that may cover a worldwide ocean.

event horizon: The poetic name given to an object's *black hole radius*: the distance from a *black hole*'s center that marks the point of no return, because nothing can escape from the black hole's *gravitational force* after passing inward through the event horizon. The event horizon may be considered to be the "edge" of a black hole.

evolution: In biology, the ongoing result of *natural selection*, which under certain circumstances causes groups of similar organisms, called species, to change over time so that their descendants differ significantly in structure and appearance; in general, any gradual change of an object into another form or state of development.

exoplanet: A *planet* that orbits a *star* other than the Sun.

extremophile: Organisms that thrive at high *temperatures*, typically between 70 and 100 degrees Celsius.

Fahrenheit temperature scale: The *temperature* scale named for the German-born physicist Daniel Gabriel Fahrenheit (1686–1736), who introduced it in 1724, according to which water freezes at 32 degrees and boils at 212 degrees.

fission: The splitting of a larger atomic *nucleus* into two or more smaller nuclei. The fission of nuclei larger than iron releases energy. This fission (also called atomic fission) is the source of energy in all present-day nuclear power plants.

force: Broadly, action that tends to produce a physical change; an influence that tends to *accelerate* an object in the direction in which the force is applied to the object.

fossil: A remnant or trace of an ancient organism.

frequency: Of *photons*, the number of oscillations or vibrations per second.

fusion: The combining of smaller *nuclei* to form larger ones. When nuclei smaller than iron fuse, energy is released. Fusion provides the primary energy source for the world's nuclear weapons, and for all stars in the universe. Also called *nuclear fusion* and *thermonuclear fusion*.

galaxy: A large group of stars, numbering from several million up to many hundred billion, held together by the stars' mutual gravitational attraction, and also usually containing significant amounts of gas and dust.

galaxy cluster: A large group of *galaxies*, usually accompanied by gas and dust and by a much greater amount of *dark matter*, held together by the mutual gravitational attraction of the material forming the galaxy cluster.

Galileo **spacecraft:** The spacecraft sent by NASA to Jupiter in 1990, which arrived in December 1995, dropped a probe into Jupiter's atmosphere, and spent the next few years in orbit around the giant planet, photographing the planet and its large satellites.

gamma rays: The highest-*energy*, highest-*frequency*, and shortest-*wavelength* type of *electromagnetic radiation*.

gene: A section of a *chromosome* that specifies, by means of the genetic code, the formation of a specific chain of *amino acids*.

general theory of relativity: Introduced in 1915 by Albert Einstein, forming the natural extension of *special relativity theory* into the domain of *accelerating* objects, this is a modern theory of gravity that successfully explains many experimental results not otherwise explainable in terms of Newton's theory of gravity. Its basic premise is the "equivalence principle," according to which a person in a spaceship, for example, cannot distinguish whether the spaceship is accelerating through space, or whether it is stationary in a gravitational field that would produce the same acceleration. From this simple yet profound principle emerges a completely reworked understanding of the nature of gravity. According to Einstein, gravity is not a *force* in the traditional meaning of the word. Gravity is the curvature of space in the vicinity of a mass. The motion of a nearby object is completely determined by its velocity and the amount of curvature that is present. As counterintuitive as this sounds, general relativity theory explains all known behavior of gravitational systems ever studied and it predicts even more counterintuitive phenomena that are continually verified by controlled experiment. For example, Einstein predicted that a strong gravity field should warp space and noticeably bend light in its vicinity. It was later shown that starlight passing near the edge of the Sun (as seen during a total solar eclipse) is found to be displaced from its expected position by an amount precisely matching Einstein's predictions. Perhaps the grandest application of the general theory of relativity involves the description of our expanding universe where all of space is curved from the collected gravity of hundreds of billions of galaxies. An important and currently unverified prediction is the existence of "gravitons"—particles that carry gravitational forces and communicate abrupt changes in a gravitational field like those expected to arise from a supernova explosion.

genetic code: The set of "letters" in *DNA* or *RNA* molecules, each of which specifies a particular *amino acid* and consists of three successive molecules like those that form the cross-links between the twin spirals of DNA molecules.

genome: The total complement of an organism's *genes*.

giant planet: A planet similar in size and composition to Jupiter, Saturn, Uranus, or Neptune, consisting of a solid core of rock and ice surrounded by thick layers of mainly *hydrogen* and *helium* gas, with a mass ranging from a dozen or so Earth masses up to many hundred times the mass of Earth.

gravitational forces: One of the four basic types of *forces*, always attractive, whose strength between any two objects varies in proportion to the product of the objects' masses, divided by the square of the distance between their centers.

gravitational lens: An object that exerts sufficient *gravitational force* on passing light rays to bend them, often focusing them to produce a brighter image than an observer would see without the gravitational lens.

gravitational radiation: *Radiation*, quite unlike *electromagnetic radiation* except for traveling at the speed of light, produced in relatively large amounts when massive objects move past one another or collide at high speeds and produce ripples in space.

greenhouse effect: The trapping of *infrared* radiation by gases in a planet's atmosphere, which raises the temperature on and immediately above the planet's surface.

GW170817: The source of gravitational radiation observed by the LIGO and VIRGO detectors on August 17, 2017, the result of the merger of two *neutron stars*, whose spectrum of *electromagnetic radiation* confirmed that neutron-star mergers can produce heavy elements in abundance.

habitable zone: The region surrounding a star within which the star's heat can maintain one or more *solvents* in a liquid state, hence a spherical shell around the star with an inner and an outer boundary.

halo: The outermost regions of a galaxy—occupying a volume much larger than the visible galaxy does—within which most of a galaxy's *dark matter* resides.

helium: The second lightest and second most abundant *element*, whose nuclei all contain two *protons* and either one or two *neutrons*. Stars generate energy through the *fusion* of *hydrogen* nuclei (*protons*) into helium nuclei.

hertz: A unit of *frequency*, corresponding to one vibration per second.

Hubble constant: The constant that appears in *Hubble's law* and relates galaxies' distances to their recession velocities.

Hubble's law: The summary of the *universe*'s expansion as observed today, which states that the recession velocities of faraway *galaxies* equals a constant times the galaxies' distances from the Milky Way.

Hubble Space Telescope (HST): The space-borne telescope launched in 1991 that has secured marvelous *visible light* images of a host of astronomical objects, owing to the fact that the telescope can observe the cosmos free from the blurring and absorbing effects inevitably produced by Earth's atmosphere.

hydrogen: The lightest and most abundant *element*, whose *nuclei* each contains one *proton* and a number of *neutrons* equal to zero, one, or two.

infrared: *Electromagnetic radiation* consisting of *photons* whose *wavelengths* are all somewhat longer, and whose *frequencies* are all somewhat lower, than those of the photons that form visible light.

initial singularity: The moment at which the expansion of the *universe* began, also called the *big bang*.

inner planets: The Sun's planets Mercury, Venus, Earth, and Mars, all of which are small, dense, and rocky in comparison to the *giant planets*.

interstellar cloud: A region of interstellar space considerably denser than average, typically spanning a diameter of several dozen *light-years*, with densities of matter that range from 10 atoms per cubic centimeter up to millions of molecules per cubic centimeter.

interstellar dust: Dust particles, each made of a million or so *atoms*, probably ejected into interstellar space from the atmospheres of highly rarefied *red-giant stars*.

interstellar gas: Gas within a *galaxy* not part of any stars.

ion: An *atom* that has lost one or more of its *electrons*.

ionization: The process of converting an *atom* into an *ion* by stripping the atom of one or more *electrons*.

irregular galaxy: A *galaxy* whose shape is irregular, that is, neither *spiral* (disklike) nor *elliptical*.

isotope: *Nuclei* of a specific *element*, all of which contain the same number of *protons* but different numbers of *neutrons*.

James Webb Space Telescope (JWST): The space-borne telescope that supersedes the *Hubble Space Telescope*, carrying a larger mirror and more advanced instruments into space.

Kelvin (absolute) temperature scale: The *temperature* scale named for Lord Kelvin (William Thomson, 1824–1907) and created during the mid-nineteenth century, for which the coldest possible temperature is, by definition, 0 degrees. The temperature intervals on this scale (denoted by K) are the same as those on the *Celsius (Centigrade) temperature scale*, so that on the Kelvin scale, water freezes at 273.16 K and boils at 373.16 K.

Kepler Space Telescope: A satellite instrument designed to detect exoplanets by the transit method, launched in 2009 and deactivated in 2018.

kilogram: A unit of mass in the metric system, consisting of 1,000 grams.

kilohertz: A unit of *frequency* that describes 1,000 vibrations or oscillations per second.

kilometer: A unit of length in the metric system, equal to 1,000 meters and approximately 0.62 miles.

kinetic energy: The *energy* that an object possesses by virtue of its motion, defined as one-half of the object's mass times the square of the object's speed. Thus a more massive object, such as a truck, has more kinetic energy than a less massive object, such as a tricycle, that moves at the same speed.

Kuiper Belt: The material in orbit around the Sun at distances extending from about 40 astronomical units (AUs; Pluto's average distance) out to several hundred AUs, almost all of which is debris left over from the Sun's *protoplanetary disk*. Pluto is one of the largest objects in the Kuiper Belt.

Large Magellanic Cloud: The larger of the two irregular satellite *galaxies* of the *Milky Way*.

latitude: On Earth, the coordinate that measures north and south by specifying the number of degrees from the Equator (0°) toward the North Pole (90° north) or the South Pole (90° south).

life: A property of matter characterized by the abilities to reproduce and to *evolve*.

light (visible light): *Electromagnetic radiation* that consists of photons whose *frequencies* and *wavelengths* fall within the band denoted as *visible light*, between *infrared* and *ultraviolet*.

light-year: The distance that light or other forms of *electromagnetic radiation* travel in one year, equal to approximately 10 trillion kilometers or 6 trillion miles.

Local Group: The name given to the two dozen or so *galaxies* in the immediate vicinity of the *Milky Way* galaxy. The Local Group includes the *Large and Small Magellanic Clouds* and the *Andromeda galaxy*.

logarithmic scale: A method for plotting data whereby tremendous ranges of numbers can fit on the same piece of paper. In official terms, the logarithmic scale increases exponentially (e.g., 1, 10, 100, 1,000, 10,000) rather than arithmetically (e.g., 1, 2, 3, 4, 5).

longitude: On Earth, the coordinate that measures east or west by specifying the number of degrees from the arbitrarily defined "prime meridian," the north-south line passing through Greenwich, England. Longitudes range from 0 to 180 degrees east or 180 degrees west of Greenwich, thus including the 360 degrees that span Earth's surface.

luminosity: The total amount of *energy* emitted each second by an object in all types of *electromagnetic radiation.*

mass: A measure of an object's material content, not to be confused with weight, which measures the amount of *gravitational force* on an object. For objects at Earth's surface, however, mass and weight vary in direct proportion.

mass extinction: An event in the history of life on Earth, in some cases as the result of a massive impact, during which a significant fraction of all species of organisms become extinct within a geologically short interval of time.

megahertz: A unit of *frequency*, equal to 1 million vibrations or oscillations per second.

metabolism: The totality of an organism's chemical processes, measured by the rate at which the organism uses *energy*. A high-metabolism animal must consume energy (food) much more frequently to sustain itself.

meteor: A luminous streak of light produced by the heating of a *meteoroid* as it passes through Earth's atmosphere.

meteorite: A *meteoroid* that survives its passage through Earth's atmosphere.

meteoroid: An object of rock or metal, or a metal-rock mixture, smaller than an *asteroid*, moving in an orbit around the Sun, part of the debris left over from the formation of the solar system or from collisions between solar-system objects.

meteor shower: A large number of *meteors* observed to radiate from a specific point on the sky, the result of Earth's crossing the orbits of a large number of *meteoroids* within a short time.

meter: The fundamental unit of length in the metric system, equal to approximately 39.37 inches.

methane (CH$_4$): Molecules that each contain one *carbon* and four *hydrogen atoms*, a potent *greenhouse* gas and a potential *solvent* for extraterrestrial life.

Milky Way: The *galaxy* that contains the Sun and approximately 300 billion other stars, as well as interstellar gas and dust and a huge amount of dark matter.

model: A mental construct, often created with the aid of pencil and paper or of high-speed computers, that represents a simplified version of reality and allows scientists to attempt to isolate and to understand the most important processes occurring in a specific situation.

modified Newtonian dynamics (MOND): A variant theory of gravity proposed by the Israeli physicist Mordehai Milgrom.

molecule: A stable grouping of two or more *atoms*.

mutation: A change in an organism's *DNA* that can be inherited by descendants of that organism.

natural selection: Differential success in reproduction among organisms of the same species, the driving force behind the *evolution* of life on Earth.

nebula: A diffuse mass of gas and dust, usually lit from within by young, highly luminous stars that have recently formed from this material.

neutrino: An *elementary particle* with no *electric charge* and a mass much smaller than an *electron*'s mass, characteristically produced or absorbed in reactions among elementary particles governed by *weak forces*.

neutron: An *elementary particle* with no *electric charge*; one of the two basic components of an atomic *nucleus*.

neutron stars: The tiny remnants (less than 20 miles in diameter) of the core of a *supernova* explosion, composed almost entirely of *neutrons* and so dense that their matter effectively crams 2,000 ocean liners into each cubic inch of space.

nitrogen: The element made up of *atoms* whose *nuclei* each have 7 *protons*, and whose different *isotopes* have nuclei with 6, 7, 8, 9, or 10 neutrons. Most nitrogen nuclei have 7 neutrons.

nuclear fusion: The joining of two *nuclei* under the influence of *strong forces*, which occurs only if the nuclei approach one another at a distance approximately the size of a proton (10^{-13} centimeter).

nucleic acid: Either *DNA* or *RNA*.

nucleotide: One of the cross-linking molecules in *DNA* and *RNA*. In DNA, the four nucleotides are adenine, cytosine, guanine, and thymine; in RNA, uracil plays the role that thymine does in DNA.

nucleus (pl. nuclei): (1) the central region of an *atom*, composed of one or more *protons* and zero or more *neutrons*. (2) The region within a *eukaryotic* cell that contains the cell's genetic material in the form of chromosomes. (3) The central region of a *galaxy*.

Oort cloud: The billions or trillions of *comets* that orbit the Sun, which formed first as the *protosun* began to contract, almost all of which move in orbits thousands or even tens of thousands of times larger than Earth's orbit.

organic: Referring to chemical compounds with *carbon atoms* as an important structural element; carbon-based molecules. Also, having properties associated with life.

organism: An object endowed with the property of being alive.

oxidation: Combination with *oxygen atoms*, typified by the rusting of metals upon exposure to oxygen in Earth's atmosphere.

oxygen: The element whose *nuclei* each have 8 *protons*, and whose different *isotopes* each have 7, 8, 9, 10, 11, or 12 neutrons in each nucleus. Most oxygen nuclei have 8 neutrons to accompany their 8 protons.

ozone (O_3): Molecules made of three *oxygen atoms*, which, at high altitudes in Earth's atmosphere, shield Earth's surface against *ultraviolet* radiation.

panspermia: The hypothesis that forms of life have transferred themselves naturally from one cosmic object to another, also called "cosmic seeding."

Perseverance: A Mars-exploring rover, complete with a helicopter (*Ingenuity*), sent to Mars in 2020.

photon: An elementary particle with no mass and no *electric charge*, capable of carrying *energy*. Streams of photons form *electromagnetic radiation* and travel through space at the speed of light, 299,792 kilometers per second.

photosynthesis: The use of *energy* in the form of *visible light* or *ultraviolet photons* to produce *carbohydrate* molecules from *carbon dioxide* and water. In some organisms, hydrogen sulfide (H_2S) plays the same role that water (H_2O) does in most photosynthesis on Earth.

Planck spacecraft: An instrument launched by the European Space Agency in 2009 and retired in 2013 that superseded the WMAP satellite in making accurate observations of the *cosmic background radiation*.

planet: An object in orbit around a star that is not another star and has a size at least as large as *Pluto*, which ranks either as the Sun's smallest planet or as a *Kuiper Belt* object too small to be a planet.

planetesimal: An object much smaller than a planet, capable of building planets through numerous mutual collisions.

plate tectonics: Slow motions of plates of the crust of Earth and similar planets.

Pluto: The innermost trans-Neptunian object, discovered in 1930 and considered a planet until a vote of the International Astronomical Union in 2006 demoted it to dwarf status.

primitive atmosphere: The original atmosphere of a planet.

prokaryote: A member of one of the three domains of life, consisting of single-celled organisms in which the genetic material does not reside within a well-defined *nucleus* of the cell. Compare with *eukaryotes*.

protein: A long-chain *molecule* made of one or more chains of *amino acids*.

proton: An *elementary particle* with one unit of positive *electric charge* found in the *nucleus* of every *atom*. The number of protons in an atom's *nucleus* defines the elemental identity of that atom. For example, the *element* that has 1 proton is *hydrogen*, the one with 2 protons is *helium*, and the element with 92 protons is uranium.

proton-proton cycle: The chain of three *nuclear fusion* reactions by which most stars fuse hydrogen nuclei (*protons*) into *helium* nuclei and convert *mass* into *energy*.

protoplanet: A planet during its later stages of formation.

protoplanetary disk: The disk of gas and dust that surrounds a *star* as it forms, from and within which individual planets may form.

protostar: A *star* in formation, contracting from a much larger cloud of gas and dust as the result of its self-gravitation.

protosun: The Sun in formation.

pulsar: An object that emits regularly spaced pulses of radio *photons* (and often of higher-energy photons as well) as the result of the rapid rotation of a *neutron star*, which produces *radiation* as charged particles accelerate in the intense magnetic field associated with the neutron star.

quantum mechanics: The description of particles' behavior at the smallest scales of size, hence of the structure of *atoms* and their interaction with other atoms and *photons*, as well as the behavior of atomic *nuclei*.

quarks: Subatomic particles that bond together strongly in triplets to form *protons* and *neutrons*.

quasar (quasi-stellar radio source): An object almost starlike in appearance, but whose *spectrum* shows a large *red shift*, as a result of the object's immense distance from the *Milky Way*.

radiation: Short for *electromagnetic radiation*. In this nuclear age, the term has also come to mean any particle or form of light that is bad for your health.

radio: *Photons* with the longest *wavelengths* and lowest *frequencies*.

radioactive decay: The process by which certain types of atomic *nuclei* spontaneously transform themselves into other types.

red-giant star: A *star* that has evolved through its main sequence phase and has begun to contract its core and expand its outer layers. The contraction induces a greater rate of *nuclear fusion*, raises the star's *luminosity*, and deposits energy in the outer layers, thereby forcing the star to grow larger.

redshift: A shift to lower *frequencies* and longer *wavelengths* in the *spectrum* of a receding *galaxy*, induced by the expanding universe.

relativity: The general term used to describe Einstein's *special theory of relativity* and *general theory of relativity.*

replication: The process by which a "parent" *DNA* molecule divides into two single strands, each of which forms a "daughter" molecule identical to the parent.

resolution: The ability of a light-collecting device such as a camera, telescope, or microscope to capture detail. Resolution is always improved with larger lenses or mirrors, but this improvement may be negated for telescopes by atmospheric blurring.

revolution: Motion around another object; for example, Earth revolves around the Sun. Revolution is often confused with *rotation.*

RNA (ribonucleic acid): A large, complex molecule, made of the same types of molecules that constitute *DNA*, which performs various important functions within living cells, including carrying the genetic messages embodied in DNA to the locations where *proteins* are assembled.

rotation: The spinning of an object on its own axis. For example, Earth rotates once every 23 hours and 56 minutes.

runaway greenhouse effect: A *greenhouse effect* that grows stronger as the heating of a planet's surface increases the rate of liquid evaporation, which in turn increases the greenhouse effect.

satellite: A relatively small object that orbits a much larger and more massive one; more precisely, both objects orbit their common center of mass, in orbits whose sizes are inversely proportional to the objects' masses.

self-gravitation: The *gravitational forces* that each part of an object exerts on all the other parts.

SETI: The search for extraterrestrial intelligence.

shooting star: A popular name for a *meteor*.

silicon: The element whose *nuclei* each have 14 *protons*, whose *atoms* form strong bonds with *oxygen* atoms to create the silicate minerals that compose 90 percent of Earth's crust.

skepticism: A questioning or doubting state of mind, which lies at the root of scientific inquiry into the cosmos.

Small Magellanic Cloud: The smaller of the two *irregular galaxies* that are satellites of our *Milky Way*.

solar system: The Sun plus the objects that orbit it, including *planets* and their *satellites, asteroids, meteoroids, comets*, and interplanetary dust.

solar wind: Particles ejected from the Sun, mostly *protons* and *electrons*, which emerge continuously from the Sun's outermost layers, but do so in especially large numbers at the time of an outburst called a solar flare.

solvent: A liquid capable of dissolving another substance; a liquid within which *atoms* and *molecules* can float and interact.

space-time: The mathematical combination of space and time that treats time as a coordinate with all the rights and privileges accorded space. It has been shown through the *special theory of relativity* that nature is most accurately described using a space-time formalism. It simply requires that all events be specified with space *and* time coordinates. The appropriate mathematics does not concern itself with the difference.

special theory of relativity: First proposed in 1905 by Albert Einstein, this provides a renewed understanding of space, time, and motion. The theory is based on two "Principles of Relativity": (1) the speed of light is constant for everyone no matter how you choose to measure it; and (2) the laws of physics are the same in every frame of reference that is either stationary or moving with constant velocity. The theory was later extended to include accelerating frames of reference in the *general theory of relativity*. It turns out that the two Principles of Relativity that Einstein assumed have been shown to be valid in every experiment ever performed. Einstein extended the relativity principles to their logical conclusions and predicted an array of unusual concepts, including:

- There is no such thing as absolute simultaneous events. What is simultaneous for one observer may have been separated in time for another observer.

- The faster you travel, the slower your time progresses relative to someone observing you.

- The faster you travel, the more massive you become, so the engines of your spaceship are less and less effective in increasing your speed.

- The faster you travel, the shorter your spaceship becomes—everything gets shorter in the direction of motion.

- At the speed of light, time stops, you have zero length, and your mass is infinite. Upon realizing the absurdity of this limiting case, Einstein concluded that you cannot reach the speed of light.

Experiments invented to test Einstein's theories have verified all of these predictions precisely. An excellent example is provided by particles that have decay "half-lives." After a predictable time, half are expected to decay into another particle. When these particles are sent to speeds near the speed of light (in particle accelerators), the half-life increases in the exact amount predicted by Einstein. They also get harder to accelerate, which implies that their effective mass has increased.

species: A particular type of organism, whose members possess similar anatomical characteristics and whose offspring are fertile.

spectrum (pl. spectra): The distribution of *photons* by *frequency* or *wavelength*, often shown as a graph that presents the number of photons at each specific frequency or wavelength.

sphere: The only solid shape for which every point on the surface has the same distance from the center.

spiral arms: The spiral features seen within the disk of a *spiral galaxy*, outlined by the youngest, hottest, most luminous stars and by giant clouds of gas and dust within which such stars have recently formed.

spiral galaxy: A *galaxy* characterized by a highly flattened disk of stars, gas, and dust, distinguished by *spiral arms* within the disk.

Spitzer or SIRTF (Space InfraRed Telescope Facility) spacecraft: An infrared-observing instrument, sent into orbit around Earth in 2003 and decommissioned in 2020.

standard candle: An astronomical object of known intrinsic brightness, useful for comparing different such objects seen at different distances.

standard ruler: the maximum distance covered by a sound wave at the *time of decoupling*, which specifies the largest span across which effects were possible. The increase in the size of the standard ruler from the cosmic expansion allows determination of the *Hubble constant*.

standard siren: A source of *gravitational radiation*, typically created by the merger of two *neutron stars*, whose intrinsic energy output can be determined from the time sequence of the waves that it produces.

star: A mass of gas held together by its *self-gravitation*, at the center of which *nuclear fusion* reactions turn energy of mass into photons and then into the *kinetic energy* of particles that heats the entire star, causing its surface to glow.

star cluster: A group of stars born at the same time and place, capable of enduring as a group for billions of years because of the stars' mutual gravitational attraction.

strong forces: One of the four basic types of *forces*, always attractive, that act between nucleons (*protons* and *neutrons*) to bind them together in atomic nuclei, but only if they approach one another within distances comparable to 10^{-13} cm.

sublimation: The transition from the solid to the gaseous state, or from gas to solid, without a passage through the liquid state.

submillimeter: *Electromagnetic radiation* with *frequencies* and *wavelengths* between those of *radio* and *infrared*.

sulfur: The element whose *nuclei* each have 16 *protons*, and which forms the fifth most abundant element on Earth, after *oxygen*, *silicon*, iron, and aluminum.

supermassive black hole: A *black hole* with more than a few hundred times the mass of the Sun.

supernova (pl. supernovae): A *star* that explodes at the end of its *nuclear-fusing* lifetime, attaining such an enormous *luminosity* for a few weeks that it can almost equal the energy output of an entire *galaxy*. Supernovae produce and distribute *elements* heavier than *hydrogen* and *helium* throughout interstellar space.

telescope (gamma, X-ray, ultraviolet, optical (visible), infrared, microwave, radio): Astrophysicists have designed special telescopes and detectors for each part of the *spectrum*. Some parts of this spectrum do not reach Earth's surface. To see the *gamma rays*, *X-rays*, *ultraviolet*, and *infrared* that are emitted by many cosmic objects, these telescopes must be lifted into orbit above the absorbing layers of Earth's atmosphere. The telescopes are of different designs but they do share three basic principles: (1) They collect *photons*. (2) They focus photons. And (3) they record the photons with some sort of detector.

temperature: The measure of the average *kinetic energy* of random motion within a group of particles. On the *absolute* or *Kelvin temperature scale*, the temperature of a gas is directly proportional to the average kinetic energy of the particles in the gas.

TESS (Transiting Exoplanet Survey Satellite): A satellite launched in 2018 into a highly elliptical orbit to search for exoplanets with the transit method.

theory of everything (TOE): A hypothetical master theory that explains all physical aspects of the universe.

thermal energy: The energy contained in an object (solid, liquid, or gaseous) by virtue of its atomic or molecular vibrations. The average *kinetic energy* of these vibrations is the official definition of temperature.

thermonuclear: Any process that pertains to the behavior of the atomic *nucleus* in the presence of high temperatures.

thermonuclear fusion: Another name for *nuclear fusion*, sometimes simply referred to as fusion.

thermophile: An organism that thrives at high temperatures, close to the boiling point of water.

tides: Bulges produced in a deformable object by the *gravitational force* from a nearby object, which arise from the fact that the nearby object exerts different amounts of force on different parts of the deformable object, since those parts have different distances from it.

time of decoupling: The cosmic era, 380,000 years after the big bang, when the expansion of the universe had reduced the energy of *photons* in the *cosmic background radiation* to the point that they could no longer prevent the formation of atoms.

transit: An astronomical term that describes the passage of one object in front of another along our line of sight.

UAPs (unidentified aerial phenomena): Objects formerly known as *UFOs*.

UFOs (unidentified flying objects): Objects seen in the skies of Earth for which a natural explanation cannot be easily assigned, revealing either a profound ignorance within the scientific community or a profound ignorance within the observer.

ultraviolet radiation: *Photons* with *frequencies* and *wavelengths* between those of *visible light* and *X-rays*.

universe: Usually taken to mean everything that exists, though in modern theories what we call the universe may prove to be only one part of a much larger "metaverse" or "multiverse."

virus: A complex of *nucleic acids* and *protein* molecules that can reproduce itself only within a "host" cell of another organism.

visible light: *Photons* whose *frequencies* and *wavelengths* correspond to those detected by human eyes, intermediate between those of *infrared* and *ultraviolet* radiation.

Voyager spacecraft: The two NASA spacecraft, named *Voyager 1* and *Voyager 2*, that were launched from Earth in 1978 and passed by Jupiter and Saturn a few years later; *Voyager 2* went on to encounter Uranus in 1986 and Neptune in 1989.

wavelength: The distance between successive wave crests or wave troughs; for *photons*, the distance that a photon travels while it oscillates once.

weak forces: One of the four basic types of *forces*, acting only among elementary particles at distances of about 10^{-13} cm or less, and responsible for the decay of certain elementary particles into other types. Recent investigations have shown that weak forces and *electromagnetic forces* are different aspects of a single *electroweak force*.

white dwarf: The core of a star that has fused *helium* into *carbon nuclei*, and therefore consists of carbon nuclei plus *electrons*, squeezed to a small diameter (about the size of Earth) and a high density (about 1 million times the density of water).

WMAP (Wilkinson Microwave Anisotropy Probe) satellite: The satellite, launched in 2001 and deactivated in 2010, that studied the *cosmic background radiation* in much greater detail than the *COBE satellite* could achieve.

X-rays: *Photons* with *frequencies* greater than those of *ultraviolet* but less than those of *gamma rays*.

FURTHER READING

Adams, Fred, and Greg Laughlin. *The Five Ages of the Universe: Inside the Physics of Eternity.* New York: Free Press, 1999.

Bartusiak, Marcia. *Einstein's Unfinished Symphony: The Story of a Gamble, Two Black Holes, and a New Age of Astronomy.* New Haven: Yale University Press, 2017.

Bryson, Bill. *A Short History of Nearly Everything.* New York: Broadway Books, 2003.

Eicher, David. *The New Cosmos: Answering Astronomy's Big Questions.* Cambridge, UK: Cambridge University Press, 2016.

Freese, Katherine. *The Cosmic Cocktail: Three Parts Dark Matter.* Princeton, NJ: Princeton University Press, 2014.

Goldsmith, Donald. *Connecting with the Cosmos: Nine Ways to Experience the Majesty and Mystery of the Universe.* Naperville, IL: Sourcebooks, 2002.

———. *Exoplanets: Hidden Worlds and the Quest for Extraterrestrial Life.* Cambridge, MA: Harvard University Press, 2019.

———. *The Hunt for Life on Mars.* New York: Dutton, 1997.

———. *Nemesis: The Death-Star and Other Theories of Mass Extinction.* New York: Walker Books, 1985.

———. *The Runaway Universe: The Race to Find the Future of the Cosmos.* Cambridge, MA: Perseus, 2000.

Gott, J. Richard. *Time Travel in Einstein's Universe: The Physical Possibilities of Travel through Time.* Boston: Houghton Mifflin, 2001.

Greene, Brian. *The Elegant Universe.* New York: W. W. Norton, 2000.

———. *The Fabric of the Cosmos: Space, Time, and the Texture of Reality.* New York: Knopf, 2003.

Grinspoon, David. *Lonely Planets: The Natural Philosophy of Alien Life.* New York: HarperCollins, 2003.

Guth, Alan. *The Inflationary Universe.* Cambridge, MA: Perseus, 1997.

Johnson, Sarah Stewart. *The Sirens of Mars: Searching for Life on Another World.* New York: Crown Publishing, 2020.

Kirshner, Robert. *The Extravagant Universe: Exploding Stars, Dark Energy, and the Accelerating Cosmos.* Princeton, NJ: Princeton University Press, 2002.

Knoll, Andrew. *Life on a Young Planet: The First Three Billion Years of Evolution on Earth.* Princeton, NJ: Princeton University Press, 2003.

Krauss, Lawrence. *A Universe from Nothing: Why There Is Something Rather Than Nothing.* New York: Atria Books, 2013.

Lemonick, Michael. *Echo of the Big Bang.* Princeton, NJ: Princeton University Press, 2003.

Panek, Richard. *The 4 Percent Universe: Dark Matter, Dark Energy, and the Race to Discover the Rest of Reality.* Boston, MA: Mariner Books, 2011.

Rees, Martin. *Before the Beginning: Our Universe and Others.* Cambridge, MA: Perseus, 1997.

———. *Just Six Numbers: The Deep Forces That Shape the Universe.* New York: Basic Books, 1999.

Roach, Mary. *Packing for Mars: The Curious Science of Life in the Void.* New York: W. W. Norton, 2011.

Schilling, Govert. *Ripples in Spacetime: Einstein, Gravitational Waves, and the Future of Astronomy.* Cambridge, MA: Harvard University Press, 2017.

Seager, Sara. *The Smallest Lights in the Universe: A Memoir.* New York: Crown, 2020.

Seife, Charles. *Alpha and Omega: The Search for the Beginning and End of the Universe.* New York: Viking, 2003.

Singh, Simon. *Big Bang: The Origin of the Universe.* New York: Harper Perennial, 2005.

Starkey, Natalie. *Catching Stardust: Comets, Asteroids, and the Birth of the Solar System.* London: Bloomsbury Sigma, 2018.

Stern, Alan, and David Grinspoon. *Chasing New Horizons: Inside the Epic First Mission to Pluto*. London: Picador, 2018.

Tasker, Elizabeth. *The Planet Factory: Exoplanets and the Search for a Second Earth*. London: Bloomsbury Sigma, 2017.

Tyson, Neil deGrasse. *Astrophysics for People in a Hurry*. New York: W. W. Norton, 2017.

———. *Death by Black Hole: And Other Cosmic Quandaries*. New York: W. W. Norton, 2007.

———. *Just Visiting This Planet: Merlin Answers More Questions about Everything under the Sun, Moon, and Stars*. New York: Main Street Books, 1998.

———. *Letters from an Astrophysicist*. New York: W. W. Norton, 2019.

———. *Merlin's Tour of the Universe: A Skywatcher's Guide to Everything from Mars and Quasars to Comets, Planets, Blue Moons, and Werewolves*. New York: Main Street Books, 1997.

———. *The Pluto Files: The Rise and Fall of America's Favorite Planet*. New York: W. W. Norton, 2009.

———. *The Sky Is Not the Limit: Adventures of an Urban Astrophysicist*. New York: Doubleday, 2000.

———. *Space Chronicles: Facing the Ultimate Frontier*. New York: W. W. Norton, 1994.

Tyson, Neil deGrasse, and Avis Lang. *Accessory to War: The Unspoken Alliance between Astrophysics and the Military*. New York: W. W. Norton, 2018.

Tyson, Neil deGrasse, Charles Liu, and Robert Irion. *One Universe: At Home in the Cosmos*. Washington, DC: Joseph Henry Press, 2000.

IMAGE CREDITS

FIRST INSERT

1. NASA, ESA, CSA, and STScI
2. ESA/Gaia/DPAC (CC BY-SA 3.0 IGO)
3. ESA/Planck
4. NASA/ESA/J. Mack (STScI)/J. Madrid (Australian Telescope National Facility)
5. Kees Scherer
6. International Gemini Observatory/NOIRLab/NSF/AURA
7. NASA; ESA; L. Bradley (Johns Hopkins University); R. Bouwens (University of California, Santa Cruz); H. Ford (Johns Hopkins University); and G. Illingworth (University of California, Santa Cruz)
8. ESA/Hubble & NASA
9. NASA/CXC/M. Weiss
10. ESA/NASA
11. Arne Henden (US Naval Observatory, Flagstaff)/Image Processed by Al Kelly
12. Hubble Image: NASA, ESA, K. Kuntz (JHU), F. Bresolin (University of Hawaii), J. Trauger (Jet Propulsion Lab), J. Mould (NOAO), Y.-H. Chu (University of Illinois, Urbana), and STScI; CFHT Image: Canada-France-Hawaii Telescope/J.-C.

Cuillandre/Coelum; NOAO Image: G. Jacoby, B. Bohannan, M. Hanna/NOAO/AURA/NSF

13. NASA/ESA, The Hubble Heritage Team and A. Riess (STScI)
14. NASA/ESA, The Hubble Key Project Team and The High-Z Supernova Search Team
15. X-ray: NASA/CXC/Caltech/P. Ogle et al.; Optical: NASA/STScI & R. Gendler; IR: NASA/JPL-Caltech; Radio: NSF/NRAO/VLA
16. ESO, NASA, & P. Anders (Göttingen University)
17. ESO
18. ESO
19. ESO/INAF-VST/OmegaCAM; Acknowledgement: A. Grado, L. Limatola/INAF-Capodimonte Observatory
20. NASA, ESA, and Z. Levay (STScI/AURA); PHAT Mosaic: NASA, ESA, J. Dalcanton, B.F. Williams, L.C. Johnson (University of Washington), the PHAT team, and R. Gendler; Ground-based Background Image of M31 © 2008 R. Gendler, used with permission
21. NASA, ESA, Hubble Heritage Project (STScI, AURA)

SECOND INSERT

1. Heywood, SARAO, I. Heywood
2. Hubble Space Telescope/ESA/NASA
3. ESA/PACS/SPIRE/Quang Nguyen Luong & Frederique Motte, HOBYS Key Program consortium
4. Hubble Space Telescope, NASA, ESA
5. ESO/I. Appenzeller, W. Seifert, O. Stahl, M. Zamani
6. Sean Andrews/American Astronomical Society
7. NASA/Ames Research Center/Natalie Batalha/Wendy Stenzel
8. The Virgo collaboration/CC0 1.0
9. NASA/Chris Gunn
10. ESA
11. Matthew Vandeputte
12. ESA/Rosetta/NAVCAM – CC BY-SA IGO 3.0
13. ESA/NASA SOHO

14. NASA
15. NASA
16. NASA/JPL
17. NASA/JPL-Caltech
18. NASA/JPL-Caltech
19. NASA, JPL-Caltech, MSSS
20. NASA/JPL-Caltech/ASU/MSSS
21. NASA, JPL-Caltech, UCLA, MPS, DLR, IDA
22. NASA/JPL-Caltech/SwRI/MSSS
23. NASA, JPL-Caltech, SETI Institute, Cynthia Phillips, Marty Valenti
24. NASA, ESA, A. Simon (Goddard Space Flight Center), M.H. Wong (University of California, Berkeley), and the OPAL Team
25. NASA, ESA, JPL, SSI, Cassini Imaging Team
26. NASA, ESA, and ASI (Italian Space Agency)
27. NASA, ESA, and ASI (Italian Space Agency)
28. NASA, Johns Hopkins Univ./APL, and the Southwest Research Institute
29. Wendy Freedman/American Astronomical Society
30. Hubble Legacy Archive, ESA, NASA; Processing: Amal Biju
31. NASA, ESA, and STScI
32. NASA, ESA, and J. Lotz and the HFF Team (STScI)

INDEX

Note: Page numbers followed by *n* refer to footnotes.